IGOR I SIKORSKY
HISTORICAL ARCHIVES

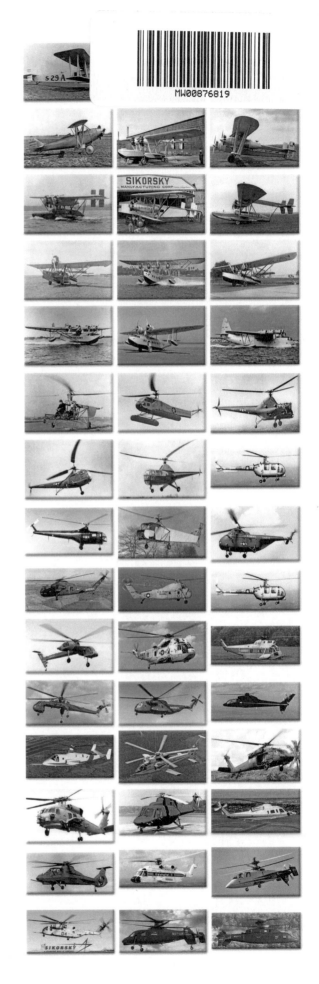

The Sikorsky Aircraft Centennial

1923-2023

A Tribute by the
Igor I. Sikorsky
Historical Archives

Frank Colucci and John Bulakowski

The Sikorsky Aircraft Centennial

1923-2023

A Tribute by the Igor I. Sikorsky Historical Archives

Frank Colucci and John Bulakowski

The Winged-S logo was created in the summer of 1928 by Mr. Sikorsky's friend, Andrei Avinoff, an artist and a director of The Pittsburg Art Museum. While vacationing with his sister's family on Long Island, N.Y., Andrei Avinoff, knowing that Igor Sikorsky had named all his aircraft with an 'S' predecessor, decided to create a logo implementing both the 'S' and a wing. He is reported to have created the Winged-S in a matter of hours and presented a large sketch of the logo to Igor Sikorsky, refusing any payment for his work. The Winged S today represents helicopter quality, safety and performance to operators around the world.

This book is dedicated to the thousands past, present and future employees responsible for the continuation of the Igor Sikorsky legacy. The fruits of their labor are demonstrated daily to the world through the performance of all the products that display the Winged S.

Foreword

In March of 1923, a small group of dedicated pioneers started an impossible enterprise. With $800.00 and several questionable promises, the Sikorsky Aero Engineering Company was born. Its founder, Igor Sikorsky, would later comment that, "it was a perfectly wrong way of entering aviation...."

Though Igor Sikorsky is best known today for his pioneering work with the helicopter, he was responsible for a number of historical aviation achievements. Back in 1913 he designed and built the world's first multi-engined aircraft, the four-engined Grand. During the 1930's he designed a series of amphibians and flying boats, including the Pan-American Clippers that established Trans-Pacific and Trans-Atlantic passenger services.

The small company struggled to stay alive from 1923 to 1927. Then Charles Lindbergh flew from New York to Paris, and suddenly America became air-minded. Igor Sikorsky was quoted as saying, "before Lindbergh's flight, aviation was a hobby, after Lindbergh's flight, aviation became a profession". As airlines formed, Sikorsky was ready with the S-38, a highly successful aircraft in that era.

In the late 1930's Igor Sikorsky returned to his first love, the helicopter. He countered the skeptics with the prediction that, "When developed, the helicopter will prove to be a unique instrument for the saving of human lives." That prediction has been proven by the millions of lives saved world-wide by the helicopter. From his backyard in Kiev, where it all started, to the current Sikorsky Aircraft Company, Igor Sikorsky would be proud. This could not have been possible without the hundreds, thousands of dedicated Sikorsky Aircraft employees who made the company what it is today.

Sergei I. Sikorsky

Sergei Sikorsky
Director Emeritus
Igor I. Sikorsky Historical Archives

Preface

This book celebrates 100 years of aviation innovation, from the start of Sikorsky Aero Engineering Corporation in 1923 to Sikorsky, a Lockheed Martin Company, today. It is the work of the Igor I. Sikorsky Historical Archives. The Archives mission is to organize, manage, and disseminate information regarding Igor Sikorsky and his legacy. Collectively, our volunteers represent 450 years of service in Sikorsky Aircraft engineering, manufacturing, program management, and other departments. We are witnesses to history.

My own history with Sikorsky Aircraft started in 1951 in the U.S. Air Force when I was one of three aircraft and engine mechanics chosen from 150 students to attend Sikorsky Helicopter School in San Marcos, Texas. We learned on the H-5E rescue helicopter, a Sikorsky S-48 with fabric-covered main rotor blades, wooden tail rotor, and bicycle chains for rotor control. I went to Vietnam in 1953 to support French forces flying Sikorsky S-55s for medical evacuation missions. As a civilian, I maintained S-55s with Petroleum Helicopters Inc. in the Gulf of Mexico flying the first offshore helicopter passenger operations. I joined Sikorsky Aircraft in 1956 and over 41 years helped introduce six Sikorsky helicopter models in the U.S. military and nine foreign countries.

My last job at Sikorsky was working on logistics proposals for Canada's S-92s – big, twin-turboshaft helicopters with composite main and tail rotors, fly-by-wire flight controls, and integrated digital cockpits. When I retired, I got involved with the Igor I. Sikorsky Historical Archives. The Archives was founded in 1995 as a 501(c)(3) non-profit organization with the approval of then-Sikorsky President Eugene Buckley. We used a basement and trailer next to the Stratford plant to preserve and display our photographs, engineering drawings, correspondence, and memorabilia, including fabric skin from the Sikorsky's pioneering helicopter, the VS-300.

The Archives has been recognized as "one of the most important resources on the history of American aviation and technology in North America" and a "National Treasure" by former curators of The Smithsonian Institution and Yale University, respectively. The wealth of volunteer experience enables us to provide in-depth research on virtually any topic related to Igor Sikorsky and Sikorsky Aircraft history.

This book is our work. It covers Igor Sikorsky's innovation in America from fixed-wing airplanes to seaplanes and amphibians to helicopters. I invite you to read and enjoy the story of Sikorsky Aircraft, and I urge you to consider the genius and commitment of Igor Sikorsky and the company that bears his name.

Dan Libertino
President, Igor I. Sikorsky Historical Archives

Introduction

The 100th anniversary of Sikorsky Aircraft is a tribute to Igor Sikorsky, the company's past and current employees, and the products that they created. This book is a history of the most prestigious vertical lift company in the world and its founder, Igor Ivanovich Sikorsky. In a 1967 interview, Mr. Sikorsky said, "The work of the individual still remains a spark which moves mankind ahead. Teamwork comes into existence after the spark." Igor Sikorsky was the spark that challenged and inspired Sikorsky employees to produce the world's finest aircraft.

Sikorsky Aircraft has employed thousands of skilled, talented, dedicated people through the years. I joined Sikorsky in 1970, and I'm privileged to have witnessed over a half-century of the company's technical and business achievements. This book takes a proud look at all 100 years. Most of it focuses on the development of the helicopter. Igor Sikorsky once observed, "The jet may have made the world smaller. The helicopter made it bigger by allowing mankind to live and work in areas that would have been inaccessible by any other vehicle." Throughout this book the headers will contain quotes. All these quotes are attributable to Igor Sikorsky.

Chapter 1 – Igor Sikorsky

Igor Sikorsky was a visionary engineer, inventor, and aviation pioneer. His life experiences, personality, compassion, and spiritual beliefs underscored his drive. In 1908, at the age of 19, he set his goal to develop an aircraft that could rise and land vertically, hover in one spot, and fly in any direction chosen by the pilot. He always believed that the greatest contribution of this device, the helicopter, would be to save lives.

Sikorsky came to America without wealth or connections. His oldest son Sergei recently observed, "America gave Igor Sikorsky the possibility of trying to do something impossible, such as starting his own aircraft company in the United States. It was this freedom that allowed him to try something, and if it didn't work, he'd pick up the pieces and try again. In his words, 'No man could forbid him to try again.'"

Chapter 2 – Fixed Wing Aircraft

Chapter 2 addresses the creation of Sikorsky Aero Engineering Corporation and its first 20 years, dominated by the design and production of fixed-wing aircraft. Igor Sikorsky started from scratch an aircraft company and kept it alive through the Great Depression (1929-1941). Undeterred by market setbacks, Sikorsky designed, built, and delivered 15 model types for a total of 233 landplanes, seaplanes and amphibians. The handsome S-38, a twin engine amphibian, was the company's first commercial success with 110 examples delivered. Working closely with Pan American Airlines, Sikorsky developed Clippers that established the first trans-Atlantic and trans-Pacific passenger routes.

Chapter 3 – Four Score of Helicopter Production

Chapter 3 encompasses the next 80 years, tracing Sikorsky's helicopter progress from the VS-300 to the latest X-2 technology rotorcraft. The VS-300 (S-46) demonstrator led to the first helicopter production line in the U.S. with the S-47 (R-4). It was during this period that the helicopter earned its reputation as a lifesaving machine. Each of 14 helicopter models designed, developed, and produced during

Igor Sikorsky's lifetime increased speed, lift, maneuverability, and mission capabilities. The S-47, for example, flew at 65 miles per hour. The S-61 was capable of 144 mph. Annual helicopter deliveries grew from double to triple digits. The S-55 and S-58 series aircraft combined to achieve sales of over 3,000 aircraft in their production runs.

By the mid-1970s, lessons learned from the Vietnam conflict (1954 – 1975) taught the U.S. Army that its thousands of utility helicopters could not meet the requirements of modern warfare. This recognition opened an opportunity for Sikorsky Aircraft to overcome a serious business challenge. New helicopter production was declining. Sikorsky Aircraft had to win the competition to replace the Army fleet.

The Army's visionary aircraft was the UTTAS (Utility Tactical Transport Aircraft System). Sikorsky Aircraft submitted its S-70 proposal in 1972 and in 1976 won the hard-fought competition for a projected Army requirement of 1,107 aircraft. The S-70 Black Hawk series today holds the company record for the greatest number of aircraft sold with over 5,000 delivered, and it will remain in production for the foreseeable future.

The fourth segment of Sikorsky's helicopter history (1980 – 2000) built upon the stability of S-70 manufacturing. This permitted Sikorsky to focus on Research and Development (R&D) projects whose technology would seed the future generations of helicopters. This industrial period saw major technological advances in the use of digital design and manufacturing technology. The logistic support community used digital data to collect, analyze, and work with engineering and the customer to improve the capability, reliability, and maintainability of helicopters.

Sikorsky helicopter history continued from the year 2000 to today with production of the S-65E, S-76, S-70 and the S-92 models and the introduction of the heavy-lift S-95 (Marine Corps CH-53K), with a maximum gross weight of 88,000 pounds. Challenges for additional capabilities, manufacturing improvements, supportability, and integration in helicopters, combined with Research and Development projects resulted in major advances. The Advancing Blade Concept demonstrated in the 1970s matured into the fast Raider and Defiant compound coaxial helicopters.

Chapters 4 and 5 – Customer Missions
Chapters 4 and 5 of this history start by providing details of Sikorsky's role in providing executive transport for the President of the U.S. and continue with the helicopter histories of each U.S. Military service. Over the last 80 years, the roles of helicopters have expanded greatly from Igor Sikorsky's fundamental wish to save lives to include a myriad of special missions flown with increasing efficiency and effectiveness.

Chapter 6 – Engineering Evolution and Epilog
This chapter recaps Sikorsky's 100-year transformation from hand drawn designs and operation sheets to leverage dramatic advances in digital engineering and manufacturing processes. These processes and leading-edge technology innovations achieved speed and range, flight control, and integrated logistics improvements to increase safety and decrease cost of ownership. They offer the company an exciting future.

Appendix 1 – Sikorsky Aircraft
Appendix 1 addresses the evolution of Sikorsky Aircraft, our facilities, corporate identities, management, and products. Readers may wish to start this book by reading the Flight Heritage section to

clarify the various nomenclatures that identify each of our aircraft.

The naming convention of all Sikorsky designs was initiated by Igor Sikorsky and continues today. All of his aircraft are identified by an 'S-#' designation where the 'S' stands for Sikorsky and the number is a sequential design starting with his S-1 aircraft during Igor Sikorsky's early days in Russia. Sikorsky's American legacy starts with the S-29A where the 'A' stood for America and runs through to Sikorsky's newest aircraft, the S-103 Defiant.

Only those aircraft which have achieved a first flight are in this section listed by first flight date. There are several identically-designated 'S' series aircraft in various position within this sequence. These aircraft have had configuration changes throughout their production lifetime that are so significant as to warrant a separate listing. The changes are predominately the result of mission requirements that dictated a major technological redesign of the baseline aircraft.

The U.S. Department of Defense has its own naming conventions and popular names for all aircraft that it procures. This can be confusing. The 'Sea King', 'Jolly Green Giant', 'Pelican', 'CH-3B', 'VH-3A', 'HH-3F', 'RH-3A' and others all refer to the same Sikorsky base model, the S-61. The cross-reference guide at the end of Appendix 1 helps readers understand what aircraft is being addressed in Chapters 2 to 6.

Appendix 2 – Igor Sikorsky's Works and Accomplishments

Igor Sikorsky knew well the importance of protecting his intellectual property. Our Archives collection includes a full set of his original design books, starting in 1926. Mr. Sikorsky had the foresight of documenting his works by signing and dating each of his ideas/calculations. This discipline continued with patents, starting with his first on March 19, 1935. The first section of this Appendix focuses on these and lists his 66 patents.

Igor Sikorsky's office at the Sikorsky Stratford facility remains intact. The office is essentially the same as it was upon Sikorsky's passing on October 26, 1972. Many of his documented honors and awards as well as significant books and photographs are located within the secure Sikorsky facility, and few have access. The second section of Appendix 2 lifts this veil.

This book concludes with an important message. It is a 1968 transcription of a presentation by Igor Sikorsky entitled *Recollections of a Pioneer*. It was addressed to new employees and remains an excellent synopsis of Igor Sikorsky's career, achievements, and beliefs in the mission of the helicopter. Those beliefs are an integral part of the Sikorsky Aircraft culture.

The bar has been set high, and the continuation of the next 100 years of Sikorsky is now our privilege, challenge, and responsibility

John Bulakowski
Vice President
Igor I. Sikorsky Historical Archives

Table of Contents

Chapter 1
Igor Sikorsky

Igor Sikorsky - The Man

During a lecture in the early years, Igor was questioned, "You have flown far, and you have flown very high. Have you ever seen God?" After a pause, he answered, No, but I have felt His Presence!"

Igor Sikorsky's inventive genius was complemented by his human compassion, a humble personality, and a strong spiritual approach to life.

This chapter is devoted to exploring the man behind the aviation legacy of Igor Sikorsky. Sikorsky has left a significant mark on the world as well as the Stratford, CT community, where he chose to expand his business ventures and make his home. His human attributes and philosophical

Spiritual convictions and beliefs

approach to life are revealed in his writings and commentary from his associates and the people who knew him best. Sergei Sikorsky's memory of his father and writings are reflected in this biography. The Sikorsky publications mentioned can be accessed from sikorskyarchives.com.

St. Nicholas Church, Stratford, CT

S-21

S-42

S-46 (VS-300)

Igor Sikorsky's ancestors can be traced to the reign of Peter the Great, 1672 to 1725. The first was a Russian Orthodox priest, Gregory Sikorsky (circa 1705 to 1760).

His son, also a priest, was Alexei Gregorivich Sikorsky (circa 1740 to 1785). Both were already established somewhere in southwestern Russia. Early records mention Alexeivich Sikorsky (1772 to 1862), then Alexei Ivanovich Sikorsky (1806 to 1870). Alexei served in the village church of Antonovka, in the province of Kiev. Ivan Alexeiv-ich Sikorsky (Igor Sikorsky's father) was born in Antonovka in 1842. At the age of nine, he was placed in a junior seminary in Kiev to study for the priesthood. At the age of twenty, he quit the senior seminary and passed the University of Kiev's entrance examination with such a high score that he was accepted on the spot and awarded a special scholarship. His brilliant ca-reer as a physician, pioneering psychiatrist and professor at the University of Kiev is a matter of record. Ivan Sikorsky was the first to break with

tradition by not becoming a Russian Orthodox priest. He lived to see his son become Russia's leading aircraft designer and pilot, and passed away peacefully in early 1919. Igor Sikorsky's humbleness and soft-spoken humor is apparent when he responded to questions regarding his roots, he would answer: "My family is of Russian origin. My grandfather and other ancestors from the time of Peter the Great were Russian Ortho-dox priests.

Consequently, the Russian nationality of the family must be considered as well established". In his autobiography, *"The Story of the Winged S"*, Igor Sikorsky wrote, "I was born on May 25th, 1889, in Kiev, situated in south western Russia."

Igor Sikorsky grew up surrounded by a loving family of three sisters, an older brother and well-educated parents. His mother Zinaida was a graduate medical student who opted to raise the family. They lived in a three-story home in Kiev, devoting a section of the house to a school for children with special needs.

With the guidance of his mother and father, Igor developed an interest in reading and learning about the new developments in science explora-tion as well as futuristic fiction. He was intrigued with the writings of Leonardo Da Vinci and Jules Verne. Vacations to countries in Europe exposed him to airships in Germany and aircraft pioneers in France. His advanced formal education was at the St. Petersburg Naval Academy and the Polytechnic Institute in Kiev.

Igor Sikorsky's Father, Mother, Three Sisters and Older Brother

Young Igor's interests included Da Vinci's Flying Helix and Verne's Rotor Driven Flying Machine

Igor Sikorsky's eldest son Sergei authored a special edition article in memory of his dad on the occasion of the thirteenth anniversary of his passing on October 26, 1972. In it, Sergei discusses his father's early years in Russia and the revolution that resulted in Sikorsky immigrating to the United States. Sergei summarized his father's world acclaimed achievements in Russian and United States aviation, as well as his human attributes.

Igor Sikorsky was a polite and pleasant person who impressed those who met him by his charm and old-world manners. However, beneath the surface, there were a number of hidden characteristics. First was his love of flying. It drew him to aviation at an early age, and it never lost its appeal. Those who flew with him often marveled at the ease with which he would get the "feel" of an aircraft within a few minutes. He could predict the handling qualities of many aircraft simply by looking at them and he was very seldom wrong in his judgment.

Another part of Igor can best be described as his being the "Renaissance Man". One could talk with him about science, history, astronomy, religion, or most any subject. He was deeply religious, but comfortably so, believing that there was much truth in all religions, if one knew where to look. Perhaps his good friend Charles Lindbergh best described the mixture of scientist and philosopher in Igor Sikorsky in the following words: "When I start to write about Igor, it is hard to know where to begin, for he was such a great man and his life covered such a broad expanse of the intellectual, material and spiritual worlds. I know of no man who so merged these spiritual worlds. I know of no man who so merged these worlds together or who could so move from one to the other to the benefit of all. His scientific designs gained from his spiritual awareness just as his spiritual awareness was enhanced by his

scientific knowledge, and he understood as few men do the essential relationships involved."

Charles Lindbergh's wife Anne described Igor Sikorsky's personality with the following statement: "The thing that is remarkable about Igor Sikorsky, is the great precision in his thought and speech, combined with an extraordinary soaring beyond facts. He can soar out with the mystics and come right back to the practical, to daily life and people. He never excludes people. Sometimes the religious-minded exclude people or force their beliefs on others. Igor never does."

Eugene E. Wilson was President of United Aircraft Corporation in 1956, and wrote an article for the Readers Digest Magazine titled, "The Most Unforgettable Character I've Ever Met." He states "Sikorsky goes to his plant each morning to confer with his staff, but his really important work takes place late at night, when he often sits in the dark and thinks to music. An unabashed mystic, he believes that some artists and writers possess the gift of seeing beyond the curtain of time and detecting cloudy visions of things to come. Modestly he suggests that engineers may also share the gift."

Last photo of Igor Sikorsky in his office on October 10, 1972

The words of Igor Sikorsky reflect the excitement and the romance of that pioneering era of flight. In the 1920s he started his second career in aviation, eventually designing the long range flying boats that were used to establish the first transpacific and transatlantic airline services.

Igor Sikorsky's love of aviation started as a young man experimenting with the unknowns as reflected by his brief but classic comments:

» I was always interested in flying. I dreamed about it as a small boy. However, at that time, flying was considered completely impossible.

» Aeronautics was neither an industry or even a science. Both were yet to come.

» It was an art and, I might say, a passion. Indeed, at that time (1909) it was a miracle.

» My first two machines were built between 1909 and 1910 and were helicopters. The first of these ships refused to leave the ground while the second could lift itself, but refused to lift me.

» Throughout 1910 and 1911, I followed the intensely interesting and romantic road of the early pioneers who built their machines without knowing how to build them and then climbed into the cockpits to try to fly their aircraft, without knowing how to fly.

» The next important step in my activity was the creation of the first large four engine airplane which proved to be a success. It was a unique aircraft. It took off at a speed of 60 miles an hour, cruised at 60 miles and stalled at 60 miles an hour.

» In the pioneering days, self-training, both in engineering problems and in piloting, was an important condition of success and even survival.

» During all the years this work was in progress, I did not forget my first love, the helicopter, and waited for the right moment to resume this work.

» For me, the greatest source of pride and satisfaction is the confirmation of my belief that the helicopter would prove to be a unique instrument for saving of human lives.

» There was also the comforting realization that nearly all discoveries were preceded by numerous failures.

» In the course of your work, you will from time to time encounter the situation where the facts and the theory do not coincide. In such circumstances, young gentlemen, it is my earnest advice to respect the facts.

» Truth in politics is optional. Truth in engineering is mandatory.

S-21 first four engine aircraft in the world, with Igor Sikorsky in the open nose compartment

Igor Sikorsky in his office with a model of the S-42 and a painting of the VS-44

Igor Sikorsky in the cockpit of the S-55

The classic definition of a "Renaissance Man", is one who has broad intellectual interests, and is accomplished in a wide range of fields. Igor Sikorsky definitely qualifies for this description. He was an inventor and genius in the field of aviation. He was interested in archeology and traveled the world to explore the variety of cultures in Egypt, Mexico, Greece and other locations.

His first attempt to conceive and develop a flying machine was one that truly defies gravity. With no formal aerodynamic background, he designed and built a coaxial rotor helicopter that could lift itself, but not the inventor. 100 years later, his company developed his concept and successfully broke the world speed record for a helicopter.

Igor Exploring Pyramids in Egypt by Camel

Crew of S-62 observing Leonardo Da Vinci on his flying helix

Igor Sikorsky's sea planes were a major factor in encouraging over-ocean travel via air routes to the Caribbean Islands, Bermuda, West Indies, Central and South America, across the Atlantic and Pacific Oceans, and regions which would be unreachable by other means. His contributions in the aviation field has effectively reduced the size of the world, and has brought all people closer together, resulting in a better understanding and appreciation of the differences between the various cultures.

Igor Sikorsky was an avid astronomer and studied the heavens to fathom the immensity of our physical universe. His writings indicate that he

Igor studying volcanoes in Mexico

marveled at the known characteristics of spacial bodies and invisible forces, and was mesmerized by its power and structure. He was knowledgeable about science, history, religion, mountain climbing, music and philosophy. The religious and humane characteristics of Igor Sikorsky is very apparent when you read the books he authored.

It should be noted that Igor Sikorsky lived through the horrific time period of the Russian Revolution, World War I and World War I I. During this period, the evils of Communism, Nazism and the German Holocaust, Fascism and Japanese Imperialism were being perpetrated on the world population. During this tumultuous time period, over 200 million lives were lost and the nuclear age began. It is very apparent that Igor Sikorsky's experiences during this time are reflected in his religious books, papers and lectures. His humanity and invention are reflected in the company that bears his name.

Chapter 2
American Airplanes

Land Based Aircraft
S-29A

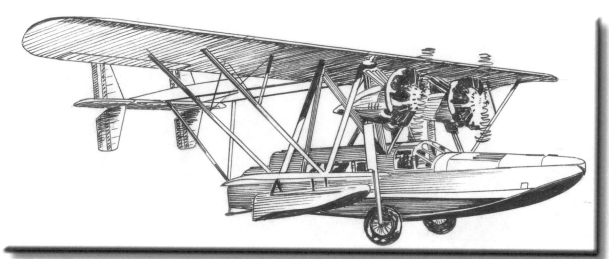

Amphibious Alicraft
S-38

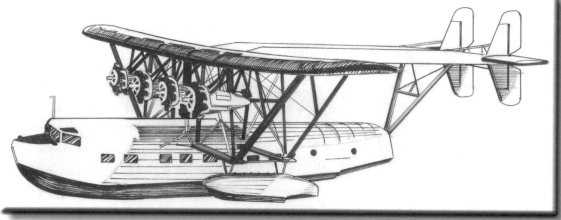

Flying Boats
S-40

Sikorsky's American Airplanes

"The successful flights of the S-42 across both major oceans may be considered as concluding the pioneering period in aviation.

While I had a good share of success and all kinds of interesting experiences, particularly with large planes and Flying Clippers, I never abandoned the idea of a helicopter."

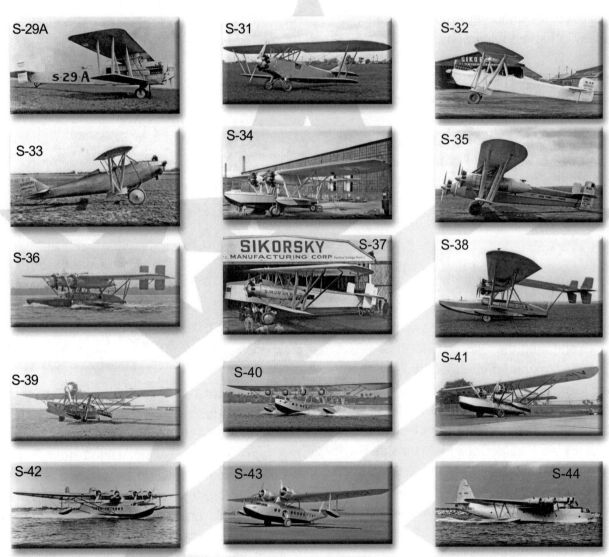

With landplanes, amphibians, and flying boats, Russian aviation pioneer Igor Sikorsky established an enduring American enterprise. (All images property Igor I. Sikorsky Historical Archives)

Igor Sikorsky arrived at Ellis Island in New York harbor on March 30, 1919 after a six-day voyage aboard the French Line steamer *La Lorraine*. Twenty-five years later, a four-engine VS-44 flying boat made by his company set a non-stop transatlantic record flying from New York LaGuardia Field to Foynes, Ireland in 14 hours, 17 minutes. With landplanes, amphibians, and flying boats, the Russian aviation pioneer established an enduring American enterprise. In his autobiography, The *Story of the Winged S*, Sikorsky recalled, "The United States seemed to me the only place which offered a real opportunity in what was then a rather precarious profession. I had been inspired by the work of Edison and Ford, the realization that a man in this country, with ideas of value -- and I hoped that mine were -- might have a chance to succeed."

Sikorsky proved himself an innovative aircraft designer and skilled aviator in Imperial Russia with the world's first successful four-engine transport, the S-21 Grand, and the S-22 to S-27 Il'ya Muramets biplane bombers of the First World War. He proposed a four-engine S-28 to be built in America, but without money or connections the émigré engineer found only temporary work at McCook Field outside Dayton, Ohio doing preliminary design on a three-engine bomber for the U.S. Army Air Service. Government jobs after World War I were few, and the director of the Army activity told Sikorsky aviation was a "dying industry."

Sikorsky lectured other Russian immigrants on mathematics, history, and astronomy in New York through 1921 and 1922. He launched Sikorsky Aero Engineering Corp. on March 5, 1923 with

money from family and friends. The company solicited subscriptions to build the all-metal, twin engine S-29A (America) passenger transport on a farm owned by Victor Utgoff near Roosevelt, Long Island. A chicken coop, shed, and garage provided crude workspaces too small for an entire aircraft, and Sikorsky recalled, "Our equipment was extremely primitive. The first instrument we devised was a home-made scissors for cutting duralumin. The material for the scissors was an old automobile bumper, bought in the neighborhood junkyard for fifty cents."

Sikorsky Aero Engineering was incorporated in New York State on March 5, 1923 and sold stock.

Sikorsky added, "That junkyard was very useful. We got from it angle irons from discarded bed springs, which proved very valuable. . . These improvisations, some of which would make a modern designer shiver with apprehension, frequently required me to redesign parts of the airplane, displacing some members to make use of the unconventional shapes which were all we could afford."

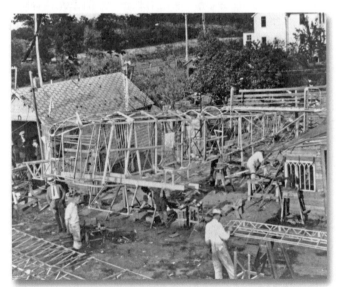

The S-29A was built by émigré volunteers at the Utgoff farm near Roosevelt, Long Island

Igor Sikorsky was president, chief engineer, and test pilot of the fledgling company. Drafts-

man Michael Gluhareff joined the enterprise in September 1924 and became chief engineer in 1925. His more efficient replacement wings for war-surplus Curtis JN4 Jenny trainers gave the company needed income. In 1924, composer Sergei Rachmaninoff contributed $5,000 that enabled Sikorsky Aero Engineering to rent leaky hangar space at nearby Roosevelt Field airport. The S-29A slid in and out of its assembly hangar sideways on reclaimed railroad tracks.

The S-29A flew at Roosevelt Field in 1924, and Sikorsky-Manufacturing Corp. was formed in 1925.

Overloaded with enthusiastic volunteer workers, the 12,000 lb, 14-seat S-29A with its second-hand Hispano Suiza engines was badly damaged on its first takeoff attempt on May 24, 1924. However, an improved S-29A with salvaged parts and overhauled Liberty engines flew on September 25, 1924 and proved itself a capable and reliable transport. Sikorsky noted his new sesquiplane (a biplane with a short-span lower wing) was one of the first twin-engine aircraft capable of sustained flight with one engine inoperative. Paying charters airlifted two grand pianos from New York to Washington DC, and S-29A flight demonstrations caught the attention of Massachusetts businessman Arnold Dickinson who invested $100,000 in 1925 ($1.58M in 2022 dollars) and became president of the reorganized Sikorsky Manufacturing Company. Igor Sikorsky remained vice president and drove new designs.

Power to Progress

The twin-engined S-30 was never finished, but Sikorsky Manufacturing Company turned out one-off examples of the single-engine ses-quiplane S-31A, S-32A, S-33 racer, and two-seat S-33A utility airplane. The five-seat S-32 with a 400 hp Liberty engine was used by an oil company in Colombia, South America. Sikorsky nevertheless concluded development of more light single-engine airplanes would be fruitless without affordable new engines. His six-seat S-34A sesquiplane in 1926 used two 225 hp Wright J-5 engines and was Sikorsky's first amphibian designed to operate from land or water. The designer explained, "It did not turn out to be practical, but it furnished valuable experience and information in a new and difficult type of multi-engined amphibians."

The S-34 was not successful but provided important lessons for later Sikorsky amphibians.

The race to cross the Atlantic non-stop meanwhile drove landplane developments. French World War I ace Renee Fonck aimed to pilot the 24,000 lb S-35A from New York to Paris but crashed the overloaded trimotor biplane on September 20, 1926. Late that year, Sikorsky moved to rented factory space in College Point, New York. For the first time he had enough room to assemble airplanes indoors. Work began on the improved S-37A for Fonck to make another trans-Atlantic attempt. Charles Lindbergh's successful crossing in 1927 negated the effort, but the 14,200 lb *Ville*

de Paris found use as a transport in Argentina. An S-37B with twin Pratt & Whitney Hornet engines was tested by the U.S. Army as a night bomber. Lindbergh's historic crossing meanwhile set the stage for a long friendship between Igor Sikorsky and the heroic American aviator that would advance commercial aviation.

The S-35 cabin could seat 12 passengers but carried extra fuel tanks for Fonck's transatlantic attempt.

The 6,000 lb S-36A amphibian with two Wright engines could be configured with standard or stretched, long-range wings. Six were built, including one for the U.S. Navy. Sikorsky acknowledged, "All this, however, was not steady work. The volume of sales was insufficient to keep the organization alive, and difficulties and worries were to a large extent, still with us. Yet the S-36 showed extremely desirable and interesting characteristics which could be achieved from a good-twin-engined amphibian. This ship, the first practical and serviceable amphibian that we produced, supplied a vast amount of experience, and we felt ready to produce a still better ship of this class."

The S-38 sesquiplane amphibian with two 420 hp Pratt & Whitney Wasp engines was completed in May 1928. Sikorsky wrote, "It was tested, and we very quickly realized that we had a really

The S-36 flown in 1927 was a bigger version of the S-34 with an enclosed passenger cabin.

excellent machine on our hands. The ship had very good takeoff characteristics from land and water. It had a climb rate of 1,000 feet per minute fully loaded, and a maximum speed close to 130 miles per hour. The ship could cruise nicely around 100 miles per hour, and it stayed in the air on one engine. All these features were excellent for 1928 and at that time there were no other amphibians with such performance characteristics."

On Feb. 4, 1929, Charles Lindbergh used the 10,800 lb S-38 to begin airmail service between Miami and the Panama Canal Zone for Pan American Airways. The airline also used Sikorsky's 10- to 12-seat aircraft to launch new mail and passenger routes to Central and South America. Private buyers placed more S-38 or-

ders. The high-performance amphibian set three International seaplane records in 1930 and, designated the XPS-1, became the first Sikorsky aircraft purchased by the U.S. Navy. Sikorsky's commercial success had a production run of 111 aircraft. The company's Winged-S logo designed by artist Andrei Avinoff debuted in an S-38 brochure.

Sikorsky Manufacturing Corp. reorganized as Sikorsky Aviation in 1929 with Igor Sikorsky as vice president of engineering. In July 1929 United Aircraft & Transport Corporation (subsequently United Aircraft Corporation and United Technologies) acquired the successful seaplane maker, and work moved to Stratford, Connecticut in a modern factory with engineering spaces, research laboratories, and a vertical wind tunnel. The five-place S-39 monoplane amphibian started as a twin-engine aircraft for executive and sport aviators but entered production in Connecticut with a single 300 hp Pratt & Whitney Wasp Jr. engine.

Working on Water

Sikorsky sold more than 20 S-39s during the Depression for around $20,000 each. Pan American Airways chose Sikorsky Manufacturing to design and build its Flying Clippers, but much bigger flying boats posed technical challenges. The 36,400 lb S-40 America Clipper was christened by U.S. First Lady Lou Hoover in October 1931.

The S-38 built in Stratford, CT found commercial success with airlines and private flyers.

The S-40 American Clipper was christened by First Lady Lou Hoover and flew PAA Caribbean routes.

Igor Sikorsky wrote in his autobiography, "While the work was in progress, serious doubts were sometimes expressed as to whether an amphibian of this size could be built. No one at that time had questioned the possibility of constructing such a seaplane or landplane, but the gross weight of 17 tons was considered too much for an amphibian, which was generally looked upon as a reasonably light airplane."

Charles Lindbergh piloted the American Clipper on its inaugural passenger-arrying flight in 1931 from Miami to Colombia and Panama with Igor Sikorsky as passenger. The designer later wrote, "Even the little experience which was made available by that time by the American Clipper was sufficient to convince me that we could build more efficient ships capable of carrying fuel for long distances, which would eventually become real equipment for transoceanic air travel." Sikorsky introduced the 12,800 lb S-41 twin-engine monoplane amphibian in 1930. In a 1935 article he wrote, "We are working hard at the present time on a flying boat of 80,000 pounds gross weight and one of 125,000 pounds gross weight, and within a few years 100- to 200-ton ships can he produced and be successful in every respect. The sizes which now are called great are to my mind not gigantic – just normal development."

The S-41 monoplane amphibian was operated by Pan American Airways and the U.S. Navy.

Ten four-engined S-42 Clippers built for Pan Am ranged in gross weight from 38,000 to 42,000 lb with 32 to 39 passenger seats. The big monoplane

flying boats set ten international records in 1934 and began Latin American service with Pan Am. In a paper for the Royal Aeronautical Society that year, Sikorsky noted, "In the past it has been intimated that efficient planes of large size would be impractical. It was argued that with increasing size the structural weight would increase beyond reasonable proportion to the gross weight, so that the ratio of useful load to gross weight would become progressively worse. It is claimed, however, that if careful study is made to distribute the power plant weight and other units of weight along the wing, stresses decrease and an efficient structure becomes possible." In 1937, Pan Am S-42Bs surveyed trans-Atlantic routes and began flying to Bermuda from New York and Baltimore.

The S-42 carried up to 32 passengers from San Francisco to Hawaii and New York to Bermuda.

Sikorsky Aircraft, Chance Vought, Hamilton Standard Propeller, and Pratt & Whitney Aircraft were consolidated as United Aircraft Manufacturing Corp. in June 1935 with Igor Sikorsky serving as division engineering manager and Michael Gluhareff as chief of design. The 15-passenger S-43 amphibian flown that year weighed just 20,000 lb but set four international performance records. Production totaled 53 aircraft, and Pan American and other airlines flew the Baby Clippers on shorter routes. The Navy and Marine Corps ordered 17 S-43/JRS-1 transports from 1937 to 1939.

The S-43 Baby Clipper flew with Pan Am, the U.S. Navy, and celebrity aviator Howard Hughes.

The Navy also ordered a single XPBS-1 (S-44) Flying Dreadnought that first flew on August 13, 1937. The four-engine, 49,000 lb patrol bomber seaplane lost its bid for a production contract, but in December 1939, American Export Airlines (AEA) gave the Vought-Sikorsky division of United Aircraft a contract for a 40-passenger version to fly transatlantic routes. The commercial VS-44A had a gross weight of 59,000 lb and then-unmatched range around 4,000 miles. It was powered by four 1,200 hp Pratt & Whitney R-1830 radial engines with Hamilton Standard propellers. AEA took deivery of the VS-44A's christened as Excalabur, Excambian and Exeter.

The first of three VS-44As for American Export Airlines was christened at Sikorsky's Stratford plant in January, 1942.

These were the only aircraft capable of flying across the North and South Atlantic nonstop with full payloads. Under a Naval Air Transport Service contract, the big transports flew from New York to Foynes, Ireland from 1942 to 1944.

Runways built around the globe during the Second World War brought an end to the big flying boats. Land-based transport aircraft were faster and more efficient. A JR2S-1 Navy transport seaplane based on the VS-44A was never built, and a wartime contract for Navy VS-44As was cancelled. A six-engine, 100- passenger flying boat proposed for Pan American Airways never took shape.

With Sikorsky Aircraft Division of United Aircraft threatened with closing, Igor Sikorsky and a small cadre of workers used the back of the VS-44A assembly area to build a new world-changing aircraft — the helicopter.

The VS-44A Excambian was restored by Sikorsky volunteers and is on display at the New England Air Museum, Windsor Locks, CT.

Chapter 3
Helicopter Legacy

Sikorsky's First Practical Helicopter
VS-300

First Production Helicopter
S-47 (R-4)

Igor Sikorsky Solves "The Helicopter Problem"

"The whole problem of the helicopter appears, therefore, interesting immediately as an engineering achievement and in the near future as a growing business proposition."

Igor Sikorsky's VS-300 demonstrator refined the world's most common helicopter configuration.

When Igor Sikorsky first flew his VS-300 helicopter on September 14, 1939, he made vertical flight discoveries that shape helicopters to this day. Most rotorcraft built since use the single main lifting rotor and single tail anti-torque rotor layout he perfected to achieve efficient hover and cruising flight. Making the helicopter safe and responsive took four years of continuous engineering, manufacturing, analysis, flight test and refinement. The talent, ingenuity, and persistence of Sikorsky and his team ultimately sustained the company that carries his name. The VS-300 legacy still serves mankind. For example, in 2022, Sikorsky S-70 Jayhawks in the U.S. Coast Guard logged their 11,000th career 'save'.

Rotary wings for vertical takeoff intrigued young Igor Sikorsky in his native Russia. However, without engines of sufficient power, the H-1 and H-2 coaxial helicopters Sikorsky built in 1909 and 1910 could not fly with a pilot aboard, and the young engineer and pilot re-focused on fixed-wing aircraft.

Later, in America, as engineering manager of one branch of the Vought-Sikorsky Aircraft Division within United Aircraft Corp. (UAC), Igor Sikorsky told UAC management of his H-3 direct-lift concept in a paper called The Helicopter Problem. In 1930, he wrote, "The most important problem to be solved in order to achieve complete success and to build a directly useful machine appears to be in the question of proper stability and control."

Igor Sikorsky recognized the achievments of earlier helicopter pioneers. According to his son Sergei, "He said that it was the Focke-Achgelis Fa-61 which was in his opinion the first practical helicopter."

The German Fa-61 with side-by-side rotors was the forerunner of bigger wartime transports halted by Allied bombing. Likewise in Germany, Flettner produced a wartime synchropter over-

Igor Sikorsky considered Germany's Focke Achgelis Fa-61 the world's first practical helicopter.

taken by events. "However, the VS-300 was the machine that perfected the single-main-rotor-plus-tail-rotor configuration," explained Sergei Sikorsky. "Today, about 95% of all the helicopters built since 1939 have followed the VS-300 configuration."

Team Demonstrator

In 1931, Igor Sikorsky filed for a patent on a direct-lift aircraft. From 1932 to 1935, his team of engineers and craftsmen tested rotor blade airfoils and helicopter models in UAC wind tunnels. In 1938, they built a rotor test rig in the Stratford, Connecticut hangar then making big VS-44A

In the first phase of Sikorsky's helicopter research, his team built a simulator with three rotors driven by a five hp electric motor. From left to right, Igor Sikorsky, Michael Buivid, Bob Labensky and Michael Gluhareff.

flying boats. The team also built a piloted helicopter engineering simulator. The tilting and rotating simulator gave the team a means to evaluate pilot inputs.

The promising first-phase investigations provided the technical basis for a helicopter. With flying boat orders running out, the Sikorsky Division of UAC was near closing. Igor Sikorsky gained the support of corporate vice president Eugene Wilson for a direct lift demonstrator – the Vought-Sikorsky VS-300 (later designated Sikorsky S-46). Sergei Sikorsky acknowledged, "He knew that he had to build a successful helicopter because otherwise Sikorsky would be folded into the Chance Vought Division at another time."

Sikorsky's bid for second-phase funding in May 1939 included a request that he retain his engineers and craftsmen. "There were a lot of good, serious individuals connected," said Sergei Sikorsky. "Eight or ten men I remember as the

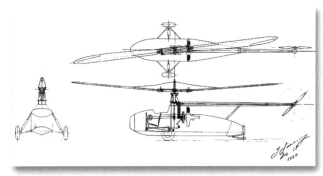

Igor Sikorsky filed a U.S. patent application in 1930 for a direct lift aircraft with a single overhead lifting rotor and a single tail anti-torque rotor.

core group of the VS-300. It was understandable they were also the core group on the prototype XR-4." The wartime XR-4 (S-47) attained production as the R-4 and HNS-1 for the U.S. Army, Coast Guard, and Navy.

Sikorsky's VS-300 team went on to work on the company's first production helicopter, the R-4. Left to right, Frank Gregory, Bob Labensky, Serge Gluhareff, Edward Walsh, Les Morris Front: Ralph Alex, Michael Buivid, George "Red" Lubben, Miles 'Bud' Roosevelt Dolph Plenefisch.

The core VS-300 team started with Igor Sikorsky and two or three Russian confidantes and grew to include others who brought proven skills to the rapid prototyping effort. Experimental engineer Boris "Bob" Labensky joined Igor Sikorsky developing the S-29 fixed-wing transport and according to Sergei Sikorsky could "build any mechanical device you needed out of spare scrap." Fellow Russian Navy and S-29 veteran George Buivid ran the experimental shop in Stratford. S-29 pilot Alex "Al" Krapish was a self-taught engineer who oversaw VS-300 assembly and maintenance.

Igor Sikorsky's cousin, Igor Alexis "Prof" Sikorsky and fellow aerodynamicist Alexander "Nick" Nikolsky led VS-300 rotor development. Brothers Serge and Michael Gluhareff filled out the senior engineering staff. Sergei Sikorsky noted, "Michael was an aerodynamic genius. . . At first, he was not too happy with the helicopter but eventually became a strong supporter." German master mechanic Adolph Plenefish built the helicopter simulator and later the VS-300. Graduate engineer George "Red" Lubben preferred the VS-300 workshop to a drawing board and was subsequently drafted into the U.S. Coast Guard where he helped design the first helicopter rescue hoists. Mechanic Henry "Hank" Wirkus went on from the VS-300 project to the Coast Guard and Army to work on early helicopters.

The first VS-300 configuration to fly was a simple, open airframe with a sheet metal tail boom. The three-bladed 28 ft diameter main rotor with full cyclic pitch control was controlled by a cyclic pitch stick on the centerline and a collective pitch wheel on the pilot's right. The main rotor blades used a symmetrical NACA airfoil built around a steel spar with spruce, mahogany, and balsa laminations. Sikorsky licensed articulated rotor head technology developed by Harold Pitcairn and inherited the low hinge offset and limited control power of the autogyro. A single-bladed, counter-balanced anti-torque rotor 40 in. in diameter commanded yaw via a single foot pedal.

A short shaft connected the main rotor to a simplified transmission and 75 hp Lycoming engine.

The first tethered VS-300 flight by Igor Sikorsky in September 1939 lasted seconds and rose inches off the ground. Sikorsky kept early untethered flights brief and maneuvers conservative. "He knew he was teaching himself to fly the helicopter at the same time he was testing the helicopter," explained Sergei Sikorsky. "He told me if something happened and the helicopter crashed, tradition blamed either the chief engineer or chief test plot. He said, 'Because I was both, I'd be held responsible no matter what happened.'"

Fly-Fix-Fly

First flight in Stratford began a continuous cycle of VS-300 discoveries and changes. Sergei Sikorsky recalled, "I would say the high point, according to my father, was that very first lift-off and two or three weeks later in December 1939 as he realized he had a machine that could fly." Years later, after a busy day at the Paris Air Show, Igor Sikorsky reflected on the first helicopter flights and confided to his son, "Vibration was fierce; control was marginal; stability was non-existent -- that was his private assessment many years later of the VS-300 in the hover." Subsequent flights by Igor Sikorsky and Serge

VS-300 aerodynamicist Serge Gluhareff shared test piloting duties with Igor Sikorsky.

Gluhareff encountered rotor control lag that caused the VS-300 to roll left as it tried to fly straight. The effect was eventually overcome with incremental changes to the rotor swashplate, but like many new helicopter challenges, answers were developed by trial and error. By November 1939, the VS-300 had a collective pitch stick to the left of the pilot and dual tail rotor pedals.

A wind gust caused the VS-300 to tip and crash on December 9, 1939, and work began on rebuilding the helicopter. In its second configuration, the VS-300 abandoned main rotor cyclic control for aircraft pitch and roll. Two horizontal tail rotors were added on outriggers to determine attitude and bank. The second configuration with its longer nose, welded-tube tail, and shock-absorbing landing gear flew on March 6, 1940. It flew before a crowd in Bridgeport that May, and Igor Sikorsky earned the first helicopter pilot's license issued by the state of Connecticut.

The agile display of the VS-300 in hover, sideways, and rearwards flight was witnessed by UAC executive vice president Eugene Wilson who later approached Igor Sikorsky and said, "Mr. Sikorsky, interesting machine. I have not seen it fly forward yet." The engineer-pilot answered, "Forward flight, Mr. Wilson, is a minor technical problem we have not solved yet." According to Sergei Sikorsky, "The problem was they underestimated the power of the lead-lag phenomena when flying forward. They had very, very primitive rubber washer-snubbers to control the fore and aft movement of the rotor blades. It was shortly after that that they put the hydraulic dampers horizontally so they would dampen the forward and the rearward motion of the rotor blades as the helicopter flew forward. That was one of the major solutions tested on the VS-300."

With successive lessons learned, the VS-300 team rebuilt their helicopter every two or three months. In July 1940, a 90 hp Franklin engine gave the VS-300 more power.

Trained by Igor Sikorsky, Army rotary wing project manager Capt. Frank Gregory flew the evolving helicopter. More flying introduced more changes. All three single-bladed tail rotors were replaced with bigger, two-bladed, hinged rotors. The horizontal tail rotor outriggers were subsequently raised 24 in. and stretched 24 in. to clear main rotor downwash. They were later inclined to tilt rotor thrust inward.

The VS-300 used two horizontal tail rotors on outriggers to provide pitch and roll control.

Igor Sikorsky received the first helicopter license from the State of Connecticut Department of Aeronautics in May 1940 after a public demonstration of the VS-300. He was again recognized with the first helicopter pilot's license from the National Aeronautic Association of the USA in December 1940.

Sikorsky VS-300
Development Configurations

Configurations

I
Sept 39 – Feb 40

First Flight – September 1939

II
Mar 40 – Jul 41

May 1940

NOSE SKID

MAIN GEAR

III
Aug 41 – Nov 41

November 1941

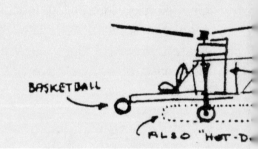

BASKETBALL

ALSO "HOT-D.

IV
Dec 41 – Oct 43

Final Flight – October 1943

SKI

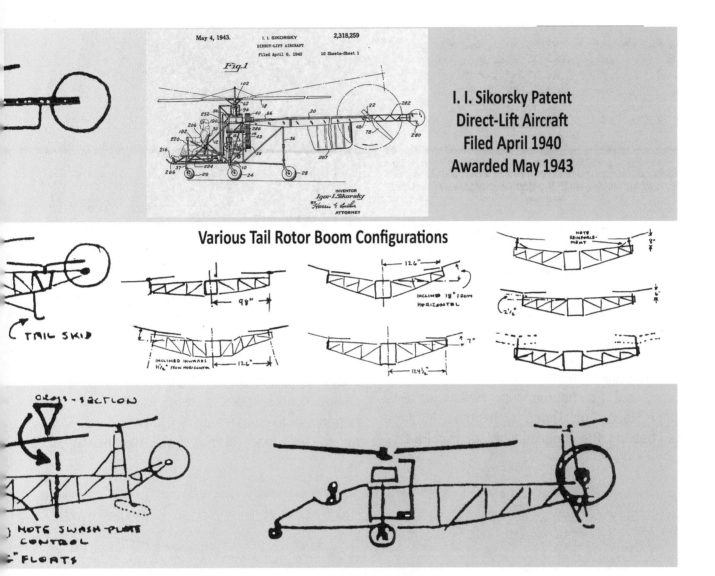

I. I. Sikorsky Patent
Direct-Lift Aircraft
Filed April 1940
Awarded May 1943

Various Tail Rotor Boom Configurations

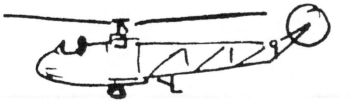

Sketches by Sergei Sikorsky

Igor Sikorsky recorded his crash in the VS-300 in October 1940, and his team rebuilt the demonstrator with lessons learned.

When an outrigger failed on October 14, 1940, the VS-300 crashed again. Igor Sikorsky emerged with minor injuries, and the team again rebuilt their helicopter.

The VS-300 in its second configuration set a U.S. Helicopter Endurance Record on April 15, 1941 when it hovered for 1 hour, 5 minutes, 14.5 seconds. Weeks later, on May 6, 1941, Igor Sikorsky set an International Helicopter Endurance Record of 1 hour, 32 minutes, 26 seconds, besting the longest flight of the Fa-61.

The VS-300 set a U.S. Helicopter Endurance Record on April 15, 1941.

The VS-300 made its first water landings with floats on the Housatonic River on April 17, 1941. It crashed again in May 1941 with Igor Sikorsky at the controls, and UAC officials made it known they no longer wanted to risk their chief helicopter engineer on test flights. Sikorsky hired Connecticut aviation commissioner and experienced fixed-wing pilot Charles "Les" Morris to test and demonstrate the helicopter.

The VS-300 with floats made its first water landings and takeoffs in April 1941 with short pontoons, tail float, and 'basketball' nose bumper. It evolved with various float configurations, culminating in the VS-300A with long "hotdog" floats that eliminated the tail float.

In June 1941, the VS-300A configuration flew with cyclic control returned to the main rotor. Sergei Sikorsky explained, "Eventually, what they wanted to do at the urging of Colonel Frank Gregory was to eliminate all of the tail rotors and stick to just one anti-torque rotor and put all cyclic control into the main rotor for a slim fuselage."

Another amphibious demonstration with pontoons installed introduced pilot Serge Gluhareff and the rest of the team to ground resonance when the shaking helicopter jumped back into the air from a ramp and dropped about 10 ft to break its tail boom.

Ground resonance caused the VS-300 to make a hard landing in June 1941.

Throughout its career the VS-300 had only rudimentary cockpit instrumentation.

Army Air Corps Captain (later Colonel) H. Franklin Gregory became the first U.S. military helicopter pilot when he flew the VS-300. As the Army helicopter project officer, Gregory wrote the VS-300 technical evaluation that led the service to commit $50,000 to the development of the XR-4. Gregory flew the first shipboard helicopter landings with the XR-4 in 1943.

On December 8, 1941, the VS-300A flew in its fourth configuration with a 30 ft diameter main rotor and single vertical tail rotor. Hydraulic dampers were added to control the lead-lag motion of the main rotor blades and eliminated serious vibration in forward flight. Sergei Sikorsky recalled, "In late December, Dad repositioned the oleos. The lead-lag vibration went away. In January, the XR-4 had oleos installed and began the Sikorsky R-4 program."

The Vought-Sikorsky VS-316, the military XR 4, with full azimuthal flight controls made its first flight on January 14, 1942, two weeks after the control problems were resolved on the VS-300A. Sergei Sikorsky observed, "When they decided to reposition the oleo dampers to horizontal lead-lag position, that was the last and final link in the engineering of the VS-300 single rotor concept." The change enabled the VS-300A to reach 80 mph. The research helicopter tried a two-bladed main rotor intermittently from October 1941 to February 1942, and while control proved satisfactory, two blades generated more vibration than three. Excessive damper loads from a three-bladed main rotor 30 ft diameter led the team to restore the original three-bladed main rotor, 28 ft diameter.

The VS-300A with a new rounded fabric nose underwent more control refinements, and drew more media attention. In the October 1942 issue

The VS-300 flew successfully with a two-bladed main rotor, albeit with higher vibration levels.

of *FLIGHTSHEET*, the publication of the Vought-Sikorsky club, editor Madeline Abell declared, "Sikorsky has produced a remarkable machine. It can rise straight up. It can descend vertically to the earth or to the water. It can climb, dive, swoop, turn, glide, dart, move sideways, forward, or backward. Which is the most astounding of its feats? The very fact that it is in existence, I think, and fortunate are those of us in contact with this, a new page in the history of aviation." Igor Sikorsky and his helicopter appeared on the cover of *LIFE* magazine in June 1943.

The VS-300 and Igor Sikorsky shared the cover of the June 21, 1943 issue of LIFE magazine, part of the growing news media recognition of the helicopter.

Also in 1943, Les Morris starred in a newsreel featuring the VS-300 in fanciful commuter service. Morris loaded groceries in the VS-300A nose basket, lifted off from a parking lot, landed ed at home in the suburbs, and flew off again from his backyard demonstrating precise control with a future commercial aircraft. The VS-300A had a maximum takeoff gross weight of 1,325 lb and could carry a 284 lb payload.

Igor Sikorsky's helicopter demonstrator was accepted enthusiastically by Sikorsky's friend Henry Ford for display in the Edison Institute in Dearborn Michigan.

Over four years of intense development, the VS-300 logged 102 hours, 34 minutes, and 51 seconds flying time. Igor Sikorsky flew his direct-lift demonstrator for the last time on October 7, 1943 before he addressed the museum crowd. The helicopter remains on display in the Heroes of the Sky exhibit in the Henry Ford Museum of American Innovation.

Demonstration pilot Les Morris flew the VS-300 in a newsreel putting the helicopter through its paces in a suburban shopping run.

Igor Sikorsky and automotive innovator Henry Ford.

Igor Sikorsky flew his VS-300 for the last time at a presentation ceremony at the Ford Museum in October, 1943.

The VS-300 remains on display in the Henry Ford Museum of American Innovation.

Four Score at Sikorsky — Part I: 1940 to 1960

"The end of the VS-300's historic career marked the start of the helicopter saga. The XR-4 became the R-4, the 'world's first production helicopter' and the only helicopter to serve in World War II. From these beginnings sprang the helicopter industry"

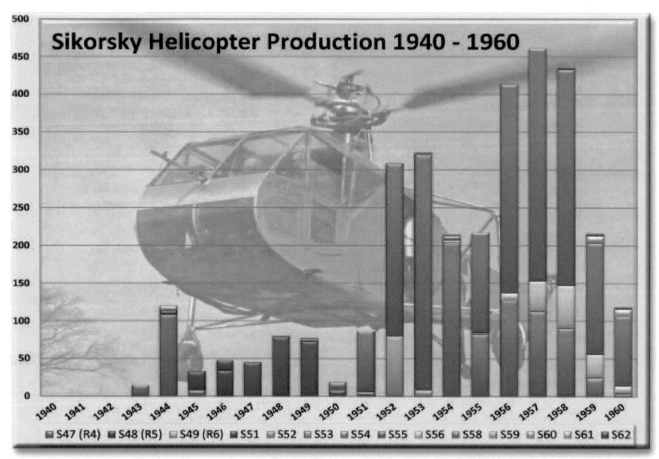

In its first 20 years of helicopter production, Sikorsky Aircraft delivered more than 3,200 helicopters.

When Igor Sikorsky flew the first public display in his VS-300 helicopter on May 20, 1940, he showed the world a truly practical means of controlled vertical flight. He also saved the company that bears his name and spurred an aviation revolution. Between 1940 and 1960, today's most common helicopter configuration – a single main lifting rotor and a tail anti-torque rotor – evolved from a shaky experiment to an enduring military and commercial success.

Soon after the VS-300 (Sikorsky S-46) debut, *TIME* magazine reported, "After work, and on Sundays, Sikorsky and helpers had puttered for months over a strange, spindle-shanked machine in a corner of United Aircraft Corp.'s Vought-Sikorsky plant, across the road from the municipal airport at Bridgeport, Conn." The Vought-Sikorsky Division then under general manager Charles McCarthy was building the big, new VS-44 flying boat on South Main Street in Stratford. However, with the flying boat market in decline, the United Aircraft board had decided to consolidate Vought-Sikorsky within Chance Vought Aircraft as of April 1, 1939.

The United Aircraft plan was to taper-off Sikorsky commercial airplane manufacture and concentrate on naval aircraft. Igor Sikorsky convinced corporate vice president Eugene Wilson to fund development of the helicopter. The VS-300 demonstrator made its first tethered flight on September 14, 1939 and with months of testing and modification showed incremental progress. The November-December 1940 issue of the United Aircraft *Bee-Hive* magazine reported, "In its ability to hover, the machine represents fulfillment of the long-awaited dream of controlled flight at zero miles per hour."

The VS-300 had powerful implications for military and commercial aviation. The *Bee-Hive* cover picture showed three brand-new Vought-Sikorsky Kingfishers, observation-scout monoplanes to be equipped with floats and catapulted from Navy battleships and cruisers. Less than a decade later, the Navy would start to replace floatplanes with helicopters. Sikorsky helicopters would replace rescue seaplanes and amphibians in the Air Force and Coast Guard. They would give Army ground forces and Marine amphibious assaults new speed, reach, and agility. In successive generations, Sikorsky helicopters would bring airmail and airlines to communities without runways, accelerate construction in remote locations, and build new businesses.

Formative Forties

The VS-300 as flown in May 1940 had three tail rotors – one vertical to counter torque and two horizontal to control pitch and roll. That July, U.S. Army Captain Frank Gregory flew the one-of-a-kind demonstrator after only a few minutes' instruction. His technical evaluation led the Army Air Corps to award a $50,000 contract to build the Vought Sikorsky VS-316 (Sikorsky S-46 or Army XR-4) as an alternative to the troubled Platt-LePage XR-1 with its big outrigger rotors. Gregory was the Project Officer for the Army Air Corps Helicopter Program and urged Igor Sikorsky to restore the single tail rotor on his evolving testbed. The VS-300A flew in its definitive configuration on Dec 8, 1941, the day after Japan attacked Pearl Harbor.

The Vought-Sikorsky Division of United Aircraft Corporation officially launched its War Production Drive in April 1942 "to increase and accelerate production and to bring home to labor and management alike the supreme importance of war production." The first employee slogan posted on Drive bulletin boards was "Fighting words are not enough; Let's get on the job and produce the stuff."

As the Stratford plant produced gull-winged Corsair fighters, the experimental XR-4 benefitted from VS-300 flight testing. A compact, controlla-

Igor Sikorsky flew his first public helicopter demonstration in 1940 with the VS-300 configured with three tail rotors.

ble helicopter with a single tail rotor could operate from ships and other confined areas.

Vought-Sikorsky presented the Army with plans for "construction of a suitable factory space capable of handling the construction of 300 helicopters per year." On June 26, 1942, Coast Guard Lt. Cmdr. Frank Erickson inspected the VS-300 and days later recommended helicopters to protect merchant convoys from submarines and rescue sailors from the sea. That July, the Navy Bureau of Aeronautics issued a Planning Directive covering procurement of four Sikorsky helicopters, a YR-4 and three XR-6s, for Navy and Coast Guard evaluation. In October, the XR-4 demonstrated the use of hydrophones to locate submarines.

With successful XR-4 tests at Wright Field in Ohio, Vought-Sikorsky received an Army contract for 15 YR-4A helicopters on December 21, 1942. A contract supplement soon added 14 more helicopters. In early 1943, Sikorsky Aircraft separated from Vought and the Stratford facility. A renovated and expanded facility on South Avenue in Bridgeport, Ct. filled with helicopter production.

In open-sea trials, British and American pilots flew the S-47 (YR-4B) from merchant ships. Bernard Whelan became general manager of United Aircraft's dedicated Sikorsky Aircraft Division in October 1943.

One YR-4 made the first helicopter combat rescue on April 25, 1944 shuttling a U.S. Army pilot and three injured British soldiers away from con-

The XR-4 demonstrated ship operations aboard the British freighter MV Daghestan in November 1943. Production R-4s flew from the ship in the Atlantic in 1944.

tested territory in Burma. Another Army helicopter successfully evacuated a wounded soldier in Burma in January 1945, but the mountain rescue underscored the performance limitations of the R-4 at high density-altitudes. The more powerful Sikorsky S-48 (Army R-5) in production through 1945 and 1946 became the first helicopter used by the Army Air Forces Air Rescue Service.

The R-5 also provided the basis for the commercial Model S-51. The S-51 first flew on February 16, 1946, and was certified a month later by the

Sikorsky helicopters evolved through the S-47, -48, and -49 (Military R-4, -5, and -6) design series, here in a simultaneous flight display at Bridgeport.

Bridgeport produced 65 S-48 (R-5) helicopters from August 1944 to October 1946. The R-5D with rescue hoist and auxiliary fuel was the first helicopter of the USAAF/USAF Air Rescue Service.

Production of the S-48 (Army R-5) at Bridgeport was cut short by the end of the Second World War, but the design evolved into the successful military and commercial S-51.

Civil Aeronautics Authority. In 1947, Los Angeles Airways initiated the world's first scheduled helicopter airmail service with five S-51's connecting Los Angeles International Airport with pickup points in a 50-mile radius. The mail routes would provide the template for helicopter passenger services.

Igor Sikorsky and Army Colonel Frank Gregory pose with an S-49 (R-6) helicopter. The Army directed wartime pro-duction of the R-6 be licensed to Nash-Kelvinator.

The start of the Cold War in 1947 drove development of more capable military helicopters. The U.S. Air Force became an independent service that year and supplemented the S-48 (Air Force H-5D) with S-51s (H-5Fs) in the Air Rescue Service. In July 1947, the Chief of Naval Operations issued a requirement for a helicopter to transport assault troops with combat equipment and supplies from escort carriers to shore. The U.S. Marine Corps used S-51 (HO3S-1) light observation helicopters to first demonstrate the concept in 1948.

The S-51 was civil-certified in 1946 and flew an airmail demonstration at Bridgeport that year. Airmail service in Los Angeles and elsewhere provided route structures for helicopter passenger service.

The December 9, 1949 issue of *Sikorsky News* featured the new 12-seat S-55 and declared, "The sleek, silver-painted helicopter, carrying an H-19 Air Force designation, made seven flights during the demonstration, and proved itself capable of taking its place in the air-world's passenger and cargo picture."

Fifties Families

Bridgeport delivered the first production H-19A to the Air Force in April 1950, and the Navy placed its first order for the S-55 (Navy HO4S-1) soon after to evaluate the helicopter in antisubmarine warfare (ASW). With the start of the Korean War in June 1950, the big-cabin helicopter found missions with all the U.S. armed services. The

Marines deployed S-55s (HRS-1s) to Korea in January 1951 with Medium Helicopter Squadron HMR-161. In their first mass troop movement, the helicopters turned a nine-hour road march into an eight-minute airlift. The Air Force sent two prototype YH-19s with an Air Proving Ground team to the combat theater in March 1951 to recover downed pilots, and production H-19s replaced H-5s in rescue squadrons over the course of the war.

The last of more than 200 S-51s came off the Bridgeport line in March 1951 for the Royal Australian Air Force, and by the end of the year, the Sikorsky factory was busy with both the three/four passenger S-52 (Marine H05S-1) and the 10-passenger S-55. The Navy established its first helicopter antisubmarine squadron, HS-1, on October 3, 1951 with S-55s playing "hunter" and "killer." Bridgeport delivered 10 S-55s in 1950, 80 in 1951, 230 in 1952, and 312 at peak production in 1953.

The versatile S-55 drew more customers. U.S. Army deployment of H-19Cs to Korea was delayed by inter-service conflict until January 1953, but by the armistice that July, the 6th Transportation Company (Helicopter) had airlifted five million pounds of supplies, 500 troops, and 1,400 sick and wounded. The Coast Guard received its first HO4S-2Gs in November 1951. Four years later, one of the helicopters drew national attention when it rescued 138 California flood victims in 29 continuous operating hours. The story helped make the helicopter the primary air rescue platform of the seaplane-centric Coast Guard.

Sikorsky realized commercial sales for the S-55 early in production. Los Angeles Airways ordered the first two civil S-55As in 1951 to begin airmail runs between Los Angeles and San Bernardino. The FAA certified the helicopter in March 1952. Belgian national airline SABENA launched the first international helicopter service on September 1, 1953. Los Angeles Airways, National Airlines, and Mohawk Airlines all started scheduled passenger services in 1954. Chicago Helicopter Airways followed suit in November 1956.

The UK Royal Navy formed 706 Squadron for anti-submarine warfare with Bridgeport-built S-55s in 1953, and Westland Helicopters began license production of the British Whirlwind soon after. License manufacture of the S-55 in France was begun by the Societe Nationale de Construction Aéronautiques Sud Est (SNCASE) in 1952. Mitsubishi in Japan started S-55 production in 1958.

The 12-seat S-55 entered production for the U.S. Air Force in 1949 and went on to equip all the U.S. serviceas, civil airlines, and international military operators.

Sikorsky had won a contract for the big, powerful S-56 (the Marine Corps HR2S-1) in May 1951 and flew the prototype XHR2S-1 at Bridgeport on December 18, 1953. The big, twin-engine helicopter quickly drew Army interest and orders. With the heavy lift helicopter still in development, the Marines and Army both ordered versions of the new Navy Seabat (Sikorsky S-58, Navy HSS-1, Marine HUS, Army CH-34). Demand for military and civil S-55s meanwhile continued. The December 23, 1954 issue of *Sikorsky News* showed an aerial view of the new factory campus being built on North Main Street in Stratford to produce the S-56, S-55, and S-58.

Navy Helicopter Antisubmarine Squadron HS-3

was the first to receive powerful new HSS-1 sub-hunter in 1955. The first Army S-58s (CH-34s) arrived at Fort Sill, Oklahoma the same year, and the Department of Defense released S-58 data for commercial sales. S-58s joined SABENA and other operators in airline service. By 1956, New York Airways S-58s were carrying 60,000 passengers a year. Automatic Stabilization Equip- ment (ASE) introduced by the Navy for hands-off night hover was certified for commercial S-58s by the Civil Aeronautics Administration. ASE also helped fly the President of the United States. The Army stood up its Presidential helicopter detachment in the fall of 1957 with four VCH-34s at Fort Belvoir, Virginia. Marine experimental squadron HMX-1 meanwhile received five similar HUS-1Zs for a Presidential Flight Detachment at Quantico. S-58 production peaked

In 1953, the S-58 (CH-34A Choctaw) gave the U.S. Army a transport helicopter powerful enough to recover downed aircraft. The S-58 went on to global military and commercial service and international production.

at 308 helicopters delivered in 1957. The 1,000th S-58, an HSS-1N, was built in the Bridgeport plant and turned over to the U.S. Navy in November 1958. License production by Westland in the UK, Fiat in Italy, and Sud Aviation in France put more of the big helicopters in service around the world.

Turbine Revolution

The S-56 that joined Army Transportation Com-

panies as the CH-37A Mojave in 1956 could carry 6,000 pounds of cargo, a Jeep with towed howitzer and gun crew, or 23 fully-equipped troops. With the record-setting HR2S-1, the Marines could airlift two combat assault squads, radio jeeps, or 5-ton sling loads from ships to shore.

The experimental S-59 (Army XH-39) set speed and altitude records in 1954 with turboshaft power. Two S-58s (Navy HSS-1Fs) started turboshaft test flights in 1957, and compact turboshafts with their high power-to-weight ratios and reduced vibration promised to increase helicopter payloads and remedy traditional helicopter drawbacks.

The 100th S-56 (a Marine HR2S) is celebrated August 28, 1958 by factory manager Alex Sperber, engineering manager Michael Gluhareff, general manager Lee Johnson, and Air Force Lt. Col. Charles Wilkins.

Igor Sikorsky announced his retirement in May 1957 but kept working in his Stratford office as an honored consultant. Long-serving general manager Bernard Whelan rose to vice president of United Aircraft that year, and Lee S. Johnson became Sikorsky general manager during a decade of turbine helicopter innovations. In an article for *Sikorsky News* Johnson noted, "While there is still some controversy among certain helicopter manufacturers and operators over the relative merits of the single main rotor versus twin main rotor design, it is interesting to note that the advocates of the twin rotor principle are fast becoming the minority. Sikorsky engineers' unswerving conviction of the merits of single-rotor designs has been supported by the apparent trend toward that configuration by manufacturers both here and abroad."

In December 1957, Sikorsky Aircraft announced development of the S-62 with S-55 dynamics driven by a single gas turbine engine. The boat-hulled helicopter flew on May 14, 1958 went into production for the Coast Guard as the HH-52.

The turbine-engined, boat-hulled S-62 demonstrated water operations in 1958 and went on to production for the Coast Guard as the HH-52A Seaguard.

Sikorsky flew the twin-turbine S-61 (Navy HSS 2) Sea King on March 11, 1959 with test pilots Robert Decker and Francis 'Yip' Yirrell. To support Sea King development, the company built at Stratford what was then the world's largest rotor test stand. The YHSS-2 prototype spawned a series of Sea King developments including shipboard sub-hunters, air-refuelable rescue

The S-61 mock-up at Bridgeport showed the twin turbine engines, boat hull, and five-bladed rotor system that would distinguish the record-breaking HSS-2 Sea King.

The S-61 Sea King flew in 1959 went into production at Stratford for the U.S. Navy and spawned a series of derivatives for the U.S. services, international civil and military operators.

helicopters, and land-based airliners that would continue in production at Stratford through 1978.

Igor Sikorsky had long forecast the value of a flying crane helicopter optimized for heavy external loads. In 1960, the Army began acquisition of a twin-turbine crane leveraging the production S-56 and the experimental S-60 Skycrane. The resulting S-64 (CH-54 Tarhe) remains in production with Erickson Aircraft today and is one candidate for Sikorsky autonomous flight technology.

The S-60 flying crane demonstrator first flew on March 25, 1959 with the rotor system and engines of the S-56 and provided the basis for the turbine-engined S-64 Skycrane.

In 20 eventful years, Sikorsky Aircraft delivered over 3,200 helicopters ranging from the exploratory VS-300 to the sophisticated Sea King and brought the power of vertical lift to users around the world.

Four Score at Sikorsky — Part II: 1960 to 1980

"This is one of the happiest days of my life. I express my deep admiration for the brilliant takes us another step forward in the history of aviation"
Igor Sikorsky during the 1967 Paris Air Show, commenting on the successful first non-stop helicopter crossing of the Atlantic Ocean.

Sikorsky Helicopter Production 1960 - 1980

Legend: S55, S58, S61, S62, S64, S65, S67, S69, S70A, S72, S76

Sikorsky produced over 2,300 helicopters between 1960 and 1980.
(All images property of Igor I. Sikorsky Historical Archives)

Sikorsky Aircraft began the 1960's busy filling orders for piston-engined S-55 and S-58 helicopters and bringing the turbine-powered S-61 and S-62 to production. Over the next 20 years, the company would give warfighters true heavy lift helicopters with the big S-64 and S-65, set new standards for battlefield and naval helicopters with the rugged S-70, and raise the bar for civil transport helicopters with the speedy S-76. Sikorsky engineers would also experiment with the coaxial rigid rotors, auxiliary thrusters, fly-by-wire controls and other innovations flying today on the Raider and Defiant high-speed compound helicopters.

The S-55 and S-58 were engineering and business successes. In April 1960, a Sikorsky-owned S-55 snagged a parachute payload to test mid-air retrieval techniques later used to recover drones and cruise missiles. Weeks later, a Marine Corps S-58 (HUS) launched a radio-guided Bullpup air-to-surface missile to demonstrate rotary-wing firepower. *Sikorsky News* in April 1960 reported delivery of the first S-55 license-built by Mitsubishi Heavy Industries in Japan and showed an S-58 crated at Stratford for shipment to the Japan Self Defense Forces. More unusual, Stratford shipped two S-58s to America's Cold War rival, the Soviet Union. When Soviet Premier Nikita Khrushchev toured Washington DC with President Eisenhower in 1959, he admired Eisenhower's executive S-58 (Marine HSS-1Z). With Presidential and State Department blessings, two helicopters were sold to the Soviets. After examination, one was cannibalized for parts and the other long flown by Soviet pilots preparing for helicopter competitions on western aircraft.

With the blessings of President Eisenhower and the U.S. State Department, Sikorsky delivered two float-equipped S-58s to America's Cold War rival, the Soviet Union.

Sikorsky Bridgeport and Stratford lines continued to deliver S-55s and S-58s to Chile, Brazil, and other military and commercial customers through the early 1960s, but piston-powered helicopters had plateaued in performance. *Sikorsky News* in April 1960 reported impending delivery

of the 154th and last production S-56 (CH-37) to the U.S. Army and noted, "Major components and the vast experience represented by the twin-engine S-56 will next appear in the S-64 Skycrane series of [turbine] helicopters." The same issue described a Massachusetts airshow where the piston-engined S-60 gave rides in its detachable 20-passenger pod. "The S-60 circled the airfield, landed, and, when the pod was detached, took off again in a demonstration of the Skycrane concept." The experimental helicopter, itself built around the S-56 engines and dynamics, had previously streamed sea minesweeping gear for the Navy.

In addition to heavy cargo and people pods, the piston powered S-60 crane demonstrator towed minesweeping gear in 1960.

'60s Workers & Warfighters

Turboshaft power promised better helicopter capability and economics. In June 1960, an S-62 with its single CT-58 gas turbine driving S-55 dynamics flew long-distance flights over the Gulf of Mexico in oil industry demonstrations. Petroleum Helicopters, Inc. became the first Gulf operator to order the new transport. S-62 orders were booked in 1961 from San Francisco-Oakland Helicopter Airlines in California, Okanagan Helicopters in Vancouver,

Canada, and Fuji Airlines in Japan. In July 1962, the S-62 became the first American turbine-powered helicopter certified by the then-Federal Aviation Agency for commercial operations. The

In July 1962, the S-62 became the first American turbine-powered helicopter certified by the then-Federal Aviation Agency for commercial operations.

The S-61L airliner made its first flight at Stratford on December 6, 1960, ordered by Los Angeles Airways to serve southern California routes in 1961.

amphibious S-62 with automatic stabilization equipment became the new search-and-rescue helicopter of the U.S. Coast Guard.

The boat-hulled S-62 also gave the Coast Guard a long-serving rescue and utility helicopter.

Production of twin-turbine S-61s (HSS-2s or SH-3As) was meanwhile underway in Stratford to replace S-58s (SH-34Js) in U.S. Navy antisubmarine warfare squadrons. Fleet deliveries began in September 1961, and the S-61 Sea King set out to make helicopter history and spawn a family of military and commercial spinoffs, including airliners. By the end of 1961, Sikorsky had received an order from Mitsubishi Heavy Industries in Japan for two stretched S-61 airliners. More S-61s were on contract for Los Angeles Airways and Chicago Helicopter Airways. The February 1962 *Sikorsky News* showed an artist's concept of the

next-generation S-65 proposed for short-haul airlines and announced, "The new turbocopter would carry 50 to 60 passengers up to 400 miles at speeds between 150 and 200 miles an hour."

On February 5, 1962, a Navy S-61 (HSS-2) broke the 15/25-kilometer world helicopter speed record flying 210.65 mph along the Connecticut shore from Milford to New Haven. In March, plans were announced to equip the joint-service Executive Flight Detachment with the S-61 (Marine Corps HSS-2Z and Army VH-3A) to transport the President of the United States. Air Force pilots began transitioning to the S-61 (CH-3B) initially ordered to support the radar warning Texas Towers off the Atlantic coast.

War in Vietnam spurred military helicopter demand and developments. On April 15, 1962, 24 S-58s (Marine UH-34Ds) of HMM-362 flew from amphibious assault ship Princeton to Soc Trang in the Republic of Vietnam for Operation Shufly, the first large unit commitment of Marines in Vietnam. Sikorsky delivered the 1,500th S-58, a Marine Seahorse, in June 1962. A month later, the company was chosen to build the new Marine heavy assault transport, the S-65 (CH-53A). *Sikorsky News* described the new six-bladed helicopter and noted, "The new aircraft will be based on the technology of the twin-turbine S-64 and will use many of the components of this earlier aircraft. . ."

Armed and armored HH-3E Jolly Green Giants gave the Air Force combat rescue helicopters to recover aircrew down in North Vietnam.

The stretched S-61R with rear cargo ramp underwecon-current testing for civil certification and Air Force qualification.

The big S-64 Skycrane put on its first public flight demonstration at Stratford on June 5, 1962 witnessed by representatives of the military, press and industry. The Skycrane with its aft-facing pilot's station was the first helicopter with a rudimentary fly-by-wire control system. *Sikorsky News* reported, "The S-64, which will lift loads as heavy as 10 tons, was developed by Sikorsky for a wide variety of military and commercial applications." The first three aircraft included one company demonstrator and two for West German licensee Vereingte FIngtechnische Werke. Sikorsky President Lee Johnson turned an S-64

The S-61R went to war in Vietnam in 1965 and evolved from the CH-3C cargo helicopter to the HH-3E combat rescue aircraft.

(YCH-54A) over to the Army Aviation Materiel Command on June 30, 1963. Army CH-54As went to Vietnam in 1965 where they recovered downed aircraft and airlifted howitzers to hilltop artillery positions.

Sikorsky shipped an S-61 (SH-3A) to Japan in August 1963 and licensed Mitsubishi Heavy Industries to build the S-61 Sea King. A 1964 contract for nine RH-3As gave the Navy its first purpose-built minesweeping helicopters. Stretched S-61s were meanwhile earning their keep in airline service. S-61Ls with Los Angeles Airways carried 171,000 passengers in 1963 alone, and LAA President Clarence Belinn subsequently called the helicopter "the counterpart of the DC-3. . . minus wings." The Air Force selected the Sikorsky-funded S-61R (CH-3C) with aft car-go ramp as its long-range rotary wing support aircraft in July 1963. By August, the company was conducting simultaneous flight tests of the commercial S-61R and Air Force CH-3C. The civilian S-61R was type-certificated by the FAA the same day the first operational helicopter was delivered to the Air Force. The Tactical Air Command deployed CH-3C cargo helicopters to Vietnam in 1965. That same year, the Air Rescue Service deployed the first armed and armored HH-3E Jolly Green Giants to Vietnam. Also in 1965, the Royal

The twin-turbine S-64 Skycrane was developed to carry heavy loads including a passenger pod for commercial and military operations.

A specialized S-61, the RH-3A, was built to stream sea-mine cable cutters and gave the Navy its first purpose-built minesweeping helicopter.

Danish Air Force placed the first European order for the S-61 Sea King. Spain took delivery of the improved SH-3D in 1966 and Westland signed a UK license production agreement in 1967. On June 1, 1967, two Jolly Green Giants of the Air Force 48th Aerospace Rescue and Recovery Squadron arrived in Paris having completed the first non-stop trans-Atlantic helicopter crossing with the help of nine in-flight refuelings. The stretched, ramped Sikorsky S-61R also won the Coast Guard competition for a Medium Range Recovery helicopter and first flew as the HH-3F on October 11, 1967.

The big, fast, six-bladed S-65 (CH-53A) flew for the first time on October 14, 1964 and provided the basis of an enduring military product line. The first Marine Sea Stallions arrived in Vietnam in January 1967. The Air Force air-refuelable HH-53B Super Jolly combat rescue helicopter first flew on March 1, 1967, and the more powerful HH-53C on June 28, 1968. The first CH-53D for the Marine Corps followed on March 6, 1969. Two CH-53D/G helicopters were delivered to Germany on September 26, 1969 to start license production. The first CH-53Ds for Israel were delivered

in 1969 and soon airlifted a Soviet air defense radar and its communications van out of Egypt across the Red Sea to Israeli-held territory.

On June 1, 1967, Igor Sikorsky met with Jolly Green Giant crews who had just completed the first non-stop helicopter crossing of the Atlantic.

Besides the vibrant production programs, Sikorsky was home to innovative developments in high-speed rotary-wing flight. The S-61F compound helicopter with wings and auxiliary jet engines reached 204 kt in 1965. Sikorsky and future parent Lockheed sub-

The prototype S-65 (CH-53A) in assembly in 1964 incorporated large forgings, chemically-milled and numerically machined structures.

mitted competing proposals for the Army Advanced Aerial Fire Support System (AAFSS) on August 11, 1965. The stillborn S-66 was a tandem-seat compound helicopter with articulated main rotor and a Rotoprop thruster that swiveled 90 degrees to play tail rotor in hover and propeller in cruise. The idea was flown but never put into production. The need for greater helicopter speed would not disappear.

The first Marine Corps S-65s (CH-53As) went to war in Vietnam in 1967, followed by Air Force HH-53Bs, and more powerful CH-53Ds and HH-53Cs.

70's Speed & Survival

When the Lockheed AAFSS failed to reach production, Sikorsky vice president John McKenna promoted the S-67 Blackhawk to fill the Army's high-speed helicopter requirement. The September 1970 *Sikorsky News* reported "more than 250 representatives of United States and foreign governments and the news media attended a flight demonstration and briefing on

By the mid-1960s, Stratford was filled with S-62 (foreground), S-61R (center) and S-65 (rear) production lines.

the new S-67 Blackhawk helicopter Sept. 22 at Sikorsky's Stratford plant." The tandem-seat demon- strator used the rotor system, drive-train and engines of the Air Force S-61R. It attained 208 kt in a shallow dive in 1970 and set a world helicopter speed record at 192 kt on a closed course in 1971. Without a customer, the S-67 program was abandoned in 1974.

S-58 production ended in 1970. Sikorsky President Wesley Kuhrt told the Helicopter Association of America that January of plans to convert piston-engined S-58s to twin-turboshaft S-58Ts. Westland in the UK developed the turbine-engined Wessex under license, but Sikorsky had moved on to the more capable S-61, S-64 and S-65. On August 24, 1970 two air-refueled Air

The S-67 Blackhawk first flew in September 1970, a company-funded attempt to satisfy a U.S. Army requirement for fast, armed helicopters.

Force S-65s (HH-53Cs) made the first non-stop trans-Pacific helicopter flight, spanning 9,000 miles from Florida to South Vietnam.

The war in Vietnam continued to make extraordinary demands on helicopters and their crews. On November 21, 1970, five HH-53s and a single HH-3E flew into the Son Tay prison camp about 20 miles west of Hanoi in a bold but unsuccessful attempt to free Americans. In 1970 and 1971, unmarked Army S-64s (CH-54s) operated routinely in Laos hauling artillery and bulldozers too heavy for the S-58s and other helicopters of Air America. In 1972 armed and armored S-61s (HH-3As) of Navy CSAR squadron HC-7 flew from ships to rescue 48 downed airmen. The Navy took delivery of the first RH-53D built for airborne mine countermeasures on October 31, 1972, but in January 1973 Operation End Sweep used borrowed Marine S-65s (CH-53As) to sweep American mines from Haiphong harbor.

On October 26, 1972, aviation pioneer and Sikorsky Aircraft founder Igor Sikorsky died at his home in Easton, Connecticut. He was 83 years old and built an enduring legacy of rotorcraft engineering, industry, and service. Igor Sikorsky started what would become Sikorsky Aircraft Division of United Aircraft Corporation on March 5, 1923. He served as Sikorsky Aircraft engineering manager until his formal retirement in 1957 at the age of 68, and he remained deeply involved with the company as an engineering consultant with regular office hours. His last day at work in his Stratford office was October 25, 1972. A funeral flyover by an S-65, two S-61s, an S-64, and S-58 honored the man and his work.

In January 1972, Erickson Aircrane became the first commercial customer for the S-64E Skycrane and by the end of the year had four of the big cranes on order. An Army CH-54B set turbine-engine helicopter time-to-climb records to 9,900 and 19,800 ft at Stratford, on April 12, 1972, but the Army bought only 37 of the improved Skycranes. In 1970 the U.S. Department of Defense (DoD) had specified a joint Army-Navy Heavy Lift Helicopter (HLH) with about twice the payload of the CH-54B. Sikorsky's three-engined, four-bladed HLH proposal lost the competition in May 1971. However, the Marine Corps broke with the HLH plan and sponsored a three-engined heavy lifter to haul twice the load of the CH-53A/D but fit the same shipboard footprint. The YCH-53E Super Stallion for the Marines flew on March 1, 1974 and the following August hovered at a gross weight of 71,700-pounds the heaviest weight ever recorded to that time by a western helicopter.

The three-engined S-65E (YCH-53E) The YCH-53E first flew on March 1, 1974 and gave the U.S. Marine Corps expanded heavy-lift capability.

For all the heavy-lift advances, major Sikorsky production programs were running out. The last HH-3F was delivered to the Coast Guard in 1972. Commercial S-61 orders were sporadic, and S-65 airliners never materialized. The U.S. Army Material Command in 1972 called for proposals to build a Utility Tactical Transport Aircraft System (UTTAS) to replace the Vietnam-era Huey. The Army wanted a squad-carrying assault helicopter able to perform at high density altitudes and hardened to survive mid- or high-intensity battlefields. Sikorsky and Boeing Vertol were chosen to build UTTAS prototypes in August 1972. The first S-70 (YUH-60A) flew on October 17, 1974, and by March 1976, Sikorsky had three prototypes in competitive testing and a fourth company-funded demonstrator aimed

The S-70 or Army YUH-60A Black Hawk won the UTTAS competition in December 1976 and remains in production with more than 4,000 aircraft delivered.

memorate ship-launched scout planes of World War II. The first SH-60B flew in December 1979.

Derived from the Black Hawk, the SH-60B flew in December 1979 and began a series of Seahawk and Naval Hawk helicopters in production today.

at civil certification and international orders. The Army named Sikorsky winner of the U TAS competition on December 23, 1976 with plans then for more than 1,100 Black Hawk helicopters. Company President Kuhrt, told workers, "This is good news for us, especially in view of the great potential of this aircraft for our long-range future." United Aircraft president Harry Gray, wired his congratulations to the Sikorsky team and noted that "We now have opportunity to lead the industry in production for the 1980s." The April 1978 *Sikorsky News* pictured Major General Story Stevens, head of the U.S. Army Aviation Research and Development Command, lowering the first production Black Hawk fuselage onto its final assembly fixture and beginning a continuous production run stretching over more than 40 years and more than 4,000 aircraft.

The Army success also spawned a parallel line of Navy S-70B helicopters marinized for shipboard operations. On September 1, 1977, a Navy competition chose Sikorsky to build the aircraft of the Light Airborne Multi-Purpose System (LAMPS Mk. III). IBM Systems Integration Division (now Sikorsky Systems Integration in Owego, New York) was awarded a full-scale development contract in 1978 to integrate the LAMPS III helicopter with ship displays via real-time datalink. On February 9, 1979, the Secretary of the Navy gave the SH-60B helicopter the name Seahawk to com-

To build its commercial portfolio, Sikorsky began development of the S-74 drawing on aerodynamic, structural, and dynamic components from the military Black Hawk. The efficient, 14-seat, twin-turbine transport was aimed at the offshore oil and executive transport markets and became the S-76 Spirit in 1975 to capitalize on the U.S. bicentennial. (The Spirit name was discarded in 1980 to pursue international markets.) The S-76 prototype first flew on March 13, 1977 and started a succession of engine, avionics, and rotor improvements through more than 875 helicopters used worldwide. Significantly, 10 countries fly the 150 kt S-76 today in the Head of State mission.

The S-76 first flown in 1977 was Sikorsky's first purpose-built commercial helicopter and set new standards for offshore operations and executive transport.

The mid-1970s saw two major rotary-wing experimental aircraft programs at Sikorsky. In January 1974, the company won the competition to build two Rotor Systems Research Aircraft (RSRA) to conduct flight research on different rotor and propulsion systems. The S-72s were based on the propulsion, drivetrain, and control systems of the production S-61 Sea King but added programmable electronic flight controls, a rotor force balance system and a variable incidence wing to fly with rotors too small to carry the aircraft alone. The S-72 had a crew ejection system that would jettison rotor blades and sequence rocket-powered ejection seats. The first S-72 RSRA flew in 1976 and was configured only for conventional helicopter flight. The second RSRA was equipped for compound flight with wings and TF34 auxiliary turbofan engines on the fuselage sides. It flew for the first time on April 10, 1978.

The S-72 Rotor System Research Aircraft (RSRA) flown in 1976 was meant to test different rotor systems and integrated propulsion concepts

Higher helicopter speeds remain desirable for military and commercial applications, but the retreating rotor blades of conventional helicopters lose lift and generate excessive vibration around 200 kt. In 1971, the U.S. Army Air Mobility Research and Development Laboratory awarded Sikorsky a contract to design and build two Advancing Blade Concept (ABC) demonstrators. Coaxial, counter-rotating rotors promised to overcome retreating blade

stall and recover power wasted on tail rotors. The first S-69 (Army XH-59A) flew in July 1973 but was damaged in testing. The second aircraft was tested extensively in high- and low-speed flight. Due to high hub drag, the S-69 with only twin turboshaft power was limited to 160 kt. In 1978, it received two Pratt & Whitney auxiliary turbojets, each producing 3,000 lb thrust. The four-engined compound helicopter attained 256 kt in a joint Army-Navy-NASA flight test program, and lessons learned shape Sikorsky X2 technologies of the S-97 Raider and SB>1 Defiant flying today.

The second S-69 (XH-59A) with two auxiliary turbojets attained 256 kt, proving the Advancing Blade Concept at the heart of today's X2 technologies.

At the start of 1980, production of the S-76 was ramping up from four aircraft in January to the planned seven aircraft a month. The commercial helicopter had already laid claim to seven world speed records, including a record time from London to Paris and return in January 1980 flown by UK offshore oil operator Bristow. S-76s built in Bridgeport were being completed in West Palm Beach, Florida, and the Sikorsky Training Center opened in West Palm Beach to school customer pilots and mechanics.

Production of the S-61 ended in Stratford mid-1980 with delivery of the last S-61N to helicopter logging operator Siller Brothers in California. Customer Neal Siller, Sikorsky president Gerald J. Tobias, and production workers marked the occasion with a turnover ceremony in the

Stratford production flight hangar. By the end of the line, Sikorsky workers had produced 123 commercial S-61Ns and 13 S-61Ls.

Military business was growing stronger. Sikorsky produced More than 70 S-70 (UH-60A) Black Hawks during 1980 for the U.S. Army, and the first S-70Bs (SH-60Bs) were in test for the U.S. Navy. The first production S-65E/CH-53E Super Stallion for the U.S. Marine Corps also made its first flight at the Stratford plant on Dec. 13, 1980 and was accepted by the U.S. Navy a few days later to begin testing and the next chapters in a proud history.

New York Airways operated three S-61N helicopters to provide visitors with a bird's eye view of the Fair. Over 80,000 passengers were carried on 10,000 flights under a great variety of weather conditions.

The first production S-65E/CH-53E Super Stallion for the U.S. Marine Corps made its first flight December 13, 1980 and began its test program soon after.

Four Score at Sikorsky — Part III: 1980 to 2000

"The Saving of life is a source of great satisfaction to me. I sincerely hope that in addition to a big success in general for yourselves, that the saving of life would remain the source of satisfaction and great interest to you who are entering the helicopter field."

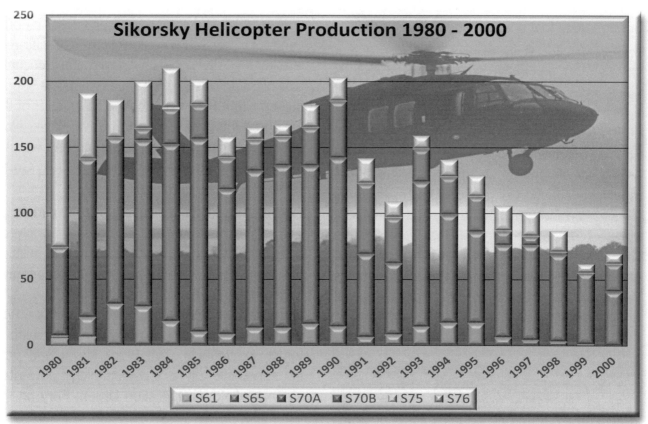

Sikorsky Helicopter Production 1980 - 2000

Legend: S61 S65 S70A S70B S75 S76

From 1980 to 2000, Sikorsky produced a range of helicopters with increasing capabilities and system sophistication.

In 1980, Sikorsky Aircraft employed 11,500 people designing, building, and supporting helicopters. The S-70 (UH-60A) Black Hawk was in full-rate production at Stratford for the U.S. Army, and the first S-70B (SH-60B) Seahawk was ready for Navy shipboard tests. The first production S-65E (CH-53E) heavy lifter for the Marine Corps underwent testing while Super Stallions two through five were in final assembly at Stratford. More than 100 S-76s had rolled off the Bridgeport line. Chief Executive Officer Robert F. Daniell told the Helicopter Association International in 1981, "Commercial helicopter operators are expressing their confidence in the Sikorsky S-76 in the best way possible -- by placing a record number of orders for them."

Daniell was named President of Sikorsky Aircraft in July 1981 as the company was evolving from an innovative aircraft maker to a complex systems integrator. The Navy was returning early-model S-61 Sea Kings to Stratford to integrate radars and other SH-3H improvements. The successor Seahawk prototype tested its advanced automatic flight control system at Stratford and went on to IBM Federal Systems (today Sikorsky Systems Integration) in Owego, New York for the Light Airborne Multi-Purpose (LAMPS) Block III mission suite that shaped today's Naval Hawks.

Special mission Black Hawks were also in test. The EH-60B with its big Standoff Target Acquisition System (SOTAS) radar first flew in February 1981. The SOTAS helicopter was never fielded, but Sikorsky later delivered EH-60A Quick Fix II Special Electronic Mission Aircraft to intercept and locate enemy radio signals. A Black Hawk medical evacuation kit passed Army operational tests at Fort Campbell, Kentucky and started the evolution of today's HH-60M Dust Off helicopter. The Army meanwhile integrated a "glass cockpit" in its UH-60A System Testbed for Avionics Research, the ancestor of today's digital cockpit Black Hawks.

In the early 1980s, the S-76, serving here with off-shore operator Air Logistics, was in demand.

Other far-reaching technologies were in the works. In early 1981, the Army Applied Technology Laboratory chose Sikorsky Aircraft as one of two for the Advanced Composite Airframe Program (ACAP) contractors to build lightweight, crashworthy, damage tolerant structures for future rotorcraft. By the end of 1981, the S-69 (Army XH-59) had achieved speeds to 263 knots exploring the Advancing Blade Concept that distinguishes today's fast Raider and Defiant compound helicopters.

'Eighties Engineering

Sikorsky sales exceeded $1 billion for the first time in 1982. The company received an Air Force contract to convert two Black Hawks into air-refuelable HH-60D Night Hawks with glass cockpits and night/adverse weather sensors. The sophisticated Night Hawk was canceled in favor of the Air Force integrated HH-60G Pave Hawk. However, it foretold the Combat Search And Rescue (CSAR) systems on today's HH-60W Jolly Green Giant II.

One ambitious helicopter program would focus Sikorsky innovations in composite structures, helicopter dynamics, flight controls, and mission systems. In late 1982, the Army documented deficiencies in its light scout/attack and utility fleets and recommended development of a new Light Helicopter Experimental (LHX) for high-intensity battlefields. The Black Hawk had proven survivable on a low-technology, low-intensity battlefield in Operation Urgent Fury on the Caribbean island of Grenada in October 1983. LHX would suppress radar, infrared, and acoustic signatures to defeat dense, integrated air defenses on future battle fields.

The S-75 ACAP demonstrator assembled in Stratford flew at West Palm Beach in 1984. It hinted at low-observable shapes for helicopters and represented the first use of Sikorsky's Computer Aided Interactive Design System for detail design and manufacturing planning. The S-75 began the engineering transition from ink-on-mylar drawings to today's digital design

The S-75 built for the Advanced Composite Airframe Program proved the advantages of fiber-reinforced composites and began the transition to digital design.

I environment. A year later, the S-76 SHADOW (initially, the Sikorsky Helicopter Advanced Demonstrator of Operator Workload) demonstrated electronic flight controls, sidearm controllers, and other cockpit innovations.

In 1984, the U.S. Army and Sikorsky had celebrated delivery of the 500th UH-60A, and the Army had awarded Sikorsky a second joint-service, multi-year production contract to deliver S-70s for the Army, Navy, and Air Force. The navalized S-70B Seahawk had been selected by Australia, Japan, and Spain, and Sikorsky announced sales of commercial S-70Cs to the People's Republic of China. A March 1985 *Sikorsky News* editorial by company president and chief executive officer William F. Paul noted 1979 Black Hawks of the first production lot each took nearly 80,000 manhours to assemble. By the eighth lot, each helicopter consumed less than 20,000 manhours.

Sikorsky used the S-76 as the basis of its SHADOW flight control and FANTAIL shrouded tail rotor technology testbeds for the LHX program.

The S-69 Advancing Blade Concept tested compound helicopter technology for today's Defiant and Raider.

Sikorsky was prime contractor for the U.S. Navy sonar-dipping SH-60F CV Helo.

Sikorsky systems integration expertise simultaneously grew more sophisticated and more valuable. In February 1985, the U.S. Navy named the company prime contractor to develop the SH-60F carrier-based anti-submarine helicopter. The so-called CV Helo would replace the Sea King. Bill Paul said, "One of the most essential CV-Helo tasks is integration of the aircraft's weapons system. Sikorsky's commitment to systems integration includes company-funded technology programs and a $38 million investment in a systems integration laboratory." Stratford opened its new, three-story research and engineering center in March 1985.

Sikorsky and Boeing announced their LHXFirst Team in June 1985 to develop a stealthy helicopter with a bearingless main rotor, fly-by-wire flight controls, night/adverse weather sensors, supercomputer processing, and digital connectivity. In 1985, the Army expected to buy 4,595 LHXs in Scout/Attack (SCAT) and Utility versions. Utility LHX succumbed to budget pressures, but the Army still wanted 1,292 SCAT helicopters for cavalry squadrons and light attack battalions. Sikorsky also broadened its commercial helicopter offerings with the more powerful S-76B certified in 1985 and demonstrated a militarized H-76

Eagle with weapons and a mast-mounted sight.

SCAT LHX promised to be "the quarterback of the digital battlefield" networked with ground and air units. Sikorsky helicopters already gave theater commanders battlespace insight. When U.S. forces clashed with Libya in the Gulf of Sidra in early 1986, Navy ships launched SH-60B Seahawks to downlink radar and electronic signatures data. When war between Iran and Iraq drew U.S. forces to the Persian Gulf in 1987 for Operation Earnest Will, Seahawks acquired thermal imagers and shared the big picture.

The first SH-60F with dipping sonar and tactical displays flew at Stratford in March 1987. "Austere" derivatives were tailored to Navy CSAR and Coast Guard Medium Range Recovery requirements. The first Navy HH-60H flew in August 1988, and *Sikorsky News* showed the first Coast Guard HH-60J on the Stratford line in March 1989. Another special mission spinoff was integrated for the President of the United States. A Stratford ceremony on November 30, 1988 marked delivery of the first of nine VH-60N White Hawks to the U.S. Marine Corps for the HMX-1 Executive Flight Detachment. Sikorsky naval helicopters were also capturing international orders. Spain received its first S-70B-1 Seahawk in 1988. Japan accepted its first

The Coast Guard HH-60J and Navy HH-60H were built on CV Helo systems integration.

S-80M-1 Sea Dragon minesweeper in 1989. Australia took its first S-70B-2 Seahawk in September 1989.

The Defense Advanced Research Projects Agency (DARPA) and NASA contracted with Sikorsky to test a four-bladed helicopter rotor that could stop in flight and become a fixed X-wing with ducted air circulation control. The intent was to have an aircraft that could hover with the efficiency of a helicopter but dash as a fixed-wing to 450 kt. The X-Wing was to be flight tested on the S-72 Rotor System Research Aircraft previously built for NASA and the Army by Sikorsky. However, funding was cut off in 1987 after much development work due to concerns about the high technical risk of a circulation-controlled stopped rotor.

Sikorsky also started work on Unmanned Air Vehicles in July 1986 under a contract from DARPA. The company used internal funds to test a shrouded coaxial rotor vehicle in July 1988. The Cypherhead proof-of-concept UAV led to a Cypher Technology Demonstrator that made its first tethered flights in April 1992. The Cypher flew tests and demonstrations through the 1990s.

'Nineties Nuances

Sikorsky moved ahead with Black Hawk improvements and international partnerships. The Army UH-60L with Seahawk improved durability gearbox and more powerful engines entered production in 1989 and set the S-70A configuration for international customers. Mitsubishi Heavy Industries in Japan signed a Black Hawk license production agreement. In 1990, Sikorsky Aircraft and Korean Air signed a license agreement to build the UH-60P Black Hawk in the Republic of Korea.

The bitter disappointment of Operation Eagle Claw in the Iranian desert a decade earlier drove the U.S. Department of Defense to fund special operations helicopters with systems for long-

The MH-60K integrated terrain avoidance radar and night-vision devices for Special Operations

range, night/adverse weather missions. Sikorsky rolled out the first MH-60K at Stratford on March 14, 1990 with integrated terrain avoidance radar, night vision devices, and a digital cockpit. Program manager Rudy Beckert remarked at the rollout, 'It carries perhaps 30% more electronics than any Black Hawk derivative we've built so far." He added, "We'll set up a line that's very similar in timing and process to what we do on the Seahawk line. This aircraft has an order-of-magnitude [greater] complexity and maybe a little more."

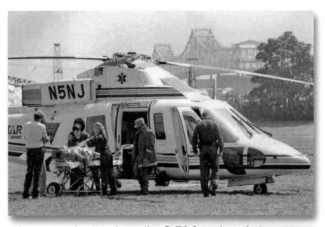

In successive versions, the S-76 found market success among aeromedical and Search-And-Rescue helicopter operators. The New Jersey State Police flew the S-76B.

Sikorsky commercial helicopters were also gaining more sophisticated and integrated avionics. In November 1990, a new S-76 search and rescue (SAR) helicopter landed on the Stratford helipad. It was the second S-76 SAR aircraft for the Royal Hong Kong Auxiliary Air Force and the first certified by the U.S. Federal Aviation Administration for hands-off approach to hover under IFR conditions.

The Army had received more than 1,000 Sikorsky Black Hawks by the time coalition forces rushed to Operation Desert Shield in 1990. In the fast-moving Desert Storm of 1991, UH-60s flew troops, equipment, supplies and the wounded. Quick Fix EH-60As located, intercepted, and jammed enemy communications. At sea, Navy SH-60Bs and SH-60Fs provided situational awareness to carrier battle groups. Marine CH-53Es and CH-53Ds hauled heavy cargo and Navy MH-53Es swept mines from sea lanes before, during, and after the war.

The YRAH-66 Comanche flew in 1996 and was the focal point of Sikorsky air vehicle and systems integration.

In April, 1991, the Boeing Sikorsky First Team won the competition to build LHX prototypes and pursue a Full Scale Development contract for what would become the RAH-66 Comanche reconnaissance attack helicopter. At an October ceremony in Stratford, Sikorsky President Eugene Buckley acknowledged, "We could not have done it without the entire company: every person

in every department and on every program we have." Plans called for the RAH-66 to be built in a joint program with Sikorsky and Boeing components and systems assembled by Sikorsky in Bridgeport. Assembly of the first prototype started in November 1993.

Sikorsky used "paperless" three-dimensioal computer-aided design for the first time on the stealthy Comanche. The design database was stored in IGOR, Sikorsky's electronic drawing vault to be updated and shared by team members in different locations. The electronic mockup provided a means to check-fit Comanche assemblies before they were built. Steel tools machined directly from CATIA files molded large composite parts that fit with precision. The first YRAH-66 flew in January 1996 with Sikorsky pilot Rus Stiles and Boeing co-pilot Bob Gradle.

Gene Buckley championed development of the next Sikorsky helicopter, and the company unveiled a wood-and-fiberglass mockup of its conceptual S-92 at the Helicopter Association International convention in 1992. The Helibus was originally a big-cabin S-61 replacement with a Black Hawk drivetrain and new-technology main rotor blades. By 1994, strict new international safety standards and tougher performance requirements from commercial operators drove development of a more powerful S-92 with damage-tolerant structures and lower operating and support costs.

Sikorsky began flying wide-chord, all-composite main rotor blades on a Black Hawk in December 1993. The risk reduction performance demonstration provided the high-lift blades for today's H-60M and S-92 helicopters. Black Hawk systems integration was also advancing the state of the cockpit art. When Sikorsky signed a contract in December 1992 for S-70s to equip the Turkish Armed Forces, it launched development of an integrated digital cockpit. In April 1996, the company announced a Black Hawk glass cockpit option and a new digital automatic flight control computer to help manage pilot workload. The improvements would provide the starting point for today's S-70M (H-60M) and S-92 crewstations.

On May 5, 1994, nearly 20 years after the first flight of the YUH-60 prototype, the 2,000th Sikorsky S-70 left Stratford with a delivery crew from the U.S. Army's 101st Aviation Brigade. A new H-60 multi-year procurement contract in 1997 covered 108 H-60s for the U.S. services including UH-60Ls for the Army, HH-60Gs for the Air Force, and 42 new CH-60 fleet combat support helicopters for the Navy. The CH-60 would evolve into today's multi-mission MH-60S and share a common cockpit with the multi-sensor MH-60R Naval Hawk.

The first S-92 flew at West Palm Beach on December 23, 1998 and marked major advances in computer aided design and manufacturing. International S-92 partners shared their computer design files in an electronic mockup, and new computer workstations enabled Sikorsky engineers to see their designs in the bigger context. Two Helibus marketing mockups were built for trade shows, but no engineering mockup was made, and partners never exchanged parts before the first aircraft was assembled. In 1999, S-92 program manager for business development Fred Geier told Vertiflite magazine, "When we assembled Aircraft One with parts from around the world, it went together smoother than our 2000th Black Hawk."

Computer aided design and advanced systems engineering would have far-reaching implications for Sikorsky business over the next two decades. The first production MH-60S made its maiden flight January 27, 2000 introducing a new generation of sophisticated, integrated Naval Hawks. Sikorsky delivered 41 new UH-60Ls that year, and the Army was making plans for a recapitalization program that would ultimately build new Black Hawks for the digital age.

The first S-92 was built by an international team in a digital engineering environment.

The MH-60S revitalized the Navy helicopter fleet and began a new series of Sikorsky Naval Hawks. (U.S. Navy)

The MH-53E minesweeper cleared sea lanes before, during, and after Desert Storm.

Four Score at Sikorsky — Part IV: 2000 to 2020

"When the time comes the helicopter would be offered for sale, the number of uses would be almost unlimited. Also inaccessible terrain would become available. We can visualize the helicopter being used for carrying mail, passengers and freight. We can further expect it in extensive use for all types of emergencies such as forest control, health, medical assistance, and developement for private use."

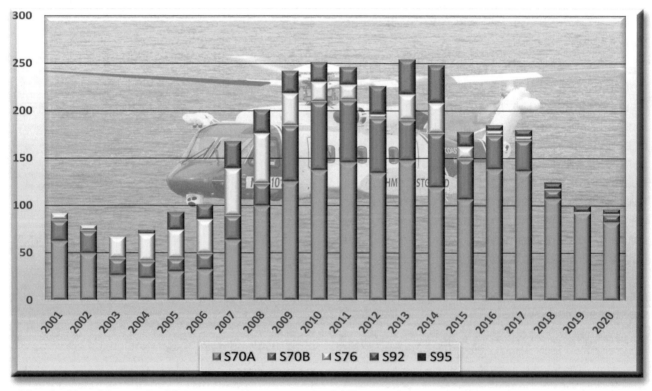

Sikorsky Aircraft began the 21st century with a strong mix of military and commercial helicopter production programs.

Sikorsky closed out the 20th century with the S-70 and S-76 in production, the S-92 in certification testing, and the RAH-66 Comanche in developmental flight test. The company was also outsourcing major components from international suppliers to be more cost-competitive in the global helicopter market. In 2000, Sikorsky chose Aero Vodochody in the Czech Republic to build S-76 airframes at lower cost than those made in Stratford. With the first S-92 order from Canadian offshore oil operator Cougar Helicopters, Stratford was integrating production structures and systems from risk-sharing partners in six countries on four continents.

The first production CH-60S marinized utility helicopter flew in January 2000 and began U.S. Navy testing that May at Patuxent River, Maryland. Though the Navy helicopter master plan aimed to modernize the fleet with new-build Knight Hawks and remanufactured SH-60R Seahawks (today, the MH-60S and MH-60R), low rate initial production was more than a year off. The U.S. Army likewise had big plans for Black Hawk remanufacture, but Stratford was building just one new S-70A (UH-60L) a month for the U.S. government under the fifth H-60 multi-year production contract. Announcing reorganization plans, company President Dean Borgman told Sikorsky employees, "The future is tremendous and it's ours to take."

The MH-60S marinized utility helicopter completed its first operational U.S. Navy deployment in May 2003.

The Boeing Sikorsky LHX team received the RAH-66 Engineering and Manufacturing Development contract in June 2000. Sikorsky's historic Bridgeport factory would be modernized for Comanche "lean" manufacturing with assembly control points networked to show workers graphic instructions on tablet computers. However, the Bridgeport airfield that launched R-4s in 1944 was not due to see low rate initial production Comanches until 2007.

To offset the U.S. Government lull, Sikorsky had healthy international and commercial business. Colombian Black Hawks were coming off the Stratford line, and marinized Thunderhawks, Seahawks and Naval Hawks were in work for Taiwan, Spain, and Turkey. New suppliers were simultaneously feeding commercial production in the U.S. The first Czech S-76 fuselage was delivered to Stratford in January 2001 for engines, dynamics, and flight certification. The 'green' S-76C+ then flew from Stratford to Keystone Helicopters in Coatesville, Pennsylvania for its corporate interior, paint, and options. Sikorsky turned Stratford warehouse space into the S-92 assembly area, and the digital design environment that linked international partners through S-92 development now orchestrated production.

Two Comanche System Design and Development helicopters were built for developmental testing.

A U.S. Army operational requirements document to recapitalize the UH-60A/L fleet was approved in March 2001, and Sikorsky received its UH-60M system design and development contract that May. The first MH-60R for the U.S. Navy – a remanufactured SH-60B -- made its maiden flight July 19, 2001 at Stratford. Terror attacks shook the world on September 11, 2001, and America's response made new demands on helicopters and Sikorsky Aircraft.

Steps at a Time

Numerous government restructurings of the Comanche program delayed fielding of the stealthy scout-attack helicopter from 2006 until 2009. More changes in the fall of 2002 cut the total buy from 1,213 to 650 Comanches at 40 odd aircraft per year and made each RAH-66 far more expensive.

Sikorsky commercial business received a major boost in December 2002 when the S-92 became the first helicopter certified under the FAA's latest and most stringent Part 29 safety amendments. The 19-passenger transport won the 2002 Collier Trophy for its safety innovations. S-76 sales meanwhile hit a 20-year high in 2003 as offshore oil operators renewed their helicopter fleets.

S-92 production used the same digital engineering environment that coordinated S-92 risk-sharing partners.

The MH-60S vertical replenishment helicopter with its marinized utility airframe and a Lockheed Martin cockpit shared by the multi-sensor MH-60R completed its first operational deployment in May 2003. That August, Sikorsky delivered the first production MH-60S configured for airborne mine countermeasures. Knight Hawks in later batches would be armed for combat rescue and anti-surface warfare. The Stratford line had also delivered prototype, test, and low rate initial production MH-60Rs to Lockheed Martin Helo Systems in Owego, New York for anti-submarine and anti-surface warfare systems.

Operation Iraqi Freedom began in March 2003,

The MH-60R Seahawk integrated dipping sonar, radar, electro-optics, and electronic support measures.

concluded "major combat operations" a month later, and settled into a long insurgency that taxed U.S. military helicopter resources, notably Army UH-60A/Ls and Marine Corps CH-53Es. The first UH-60M flew at the Development Test Center outside West Palm Beach, Florida on September 17, 2003. The next-generation M-model with its high-lift main rotor, new structures, and integrated cockpit promised to restore Black Hawk performance at high density altitudes, cut fleet costs, and make the helicopter interoperable on a digital battlefield.

The UH-60M started as a remanufacturing plan for early S-70 Black Hawks but soon gave Sikorsky an all-new production program.

Bridgeport received its first Comanche forward-fuselage for the first engineering and manufacturing development aircraft in August 2003. The RAH-66 production facility opened formally in September 2003, but the "Quarterback of the Digital Battlefield" and "Cornerstone of the New Army Vision" was by then consuming 38% of the U.S. Army aviation budget. The Army canceled the RAH-66 program altogether on February 23, 2004 with a promise to "strengthen Army Aviation" by redirecting Comanche money to broader modernization, including Black Hawk upgrades.

Also in 2004, a Joint Operational Requirements Document defined a new Marine Heavy Lift Replacement helicopter that would become the Sikorsky CH-53K. Late that year, a militarized, marinized, multi-sensor S-92 with fly-by-wire flight controls – the CH-148 Cyclone -- won the Canadian Maritime Helicopter Program.

Sikorsky acquired Schweizer Aircraft in Horseheads, New York in 2004 to provide added production capacity and a rapid prototyping "Hawk Works" similar to Lockheed Martin's 'Skunk Works'. Work on the company-funded X2 high-speed compound helicopter started in 2005. Sikorsky suffered a stinging loss in February 2005 when the U.S. Navy chose the imported EH101 over the S-92 as the U.S. Presidential Replacement Helicopter -- VXX. The Navy's ambitious and ultimately canceled two-step program aimed for a growth version of the off-the-shelf aircraft developed abroad to meet objective requirements.

Cougar Helicopters nevertheless took delivery of the first production S-92 in 2005, and Sikorsky acquired Keystone Helicopters later that year to continue S-92 completions. Black Hawk and Seahawk remanufacturing plans gave way to all-

The S-92 became the marinized, militarized CH-148 for the Canadian Maritime Helicopter Program.

new UH-60M and MH-60R production, and with Stratford full of military helicopters, Coatesville took on commercial S-92 and S-76 final assembly in stages. The S-76C++ with Ariel 2S2 engines was certified and the first aircraft delivered in December 2005.

The S-92 found notable success in offshore oil operations and introduced hands-off rig approaches.

Better and Bigger

From September 11, 2001 to the end of 2006, some 1,600 U.S. Army Black Hawks logged more than one million flight hours, nearly a third of that time in combat. Sikorsky delivered the first production UH-60M on August 3, 2006 against Army plans for 1,200 new-build Black Hawks to fly utility, Medevac, and Special Operations missions. The Marine Corps CH-53E fleet was also wearing out with extraordinary utilization. On January 2, 2006, the Navy awarded Sikorsky an initial system development and demonstration contract for what would become the CH-53K, an all-new heavy lift replacement helicopter with new-technology engines, hybrid composite structures, and fly-by-wire flight controls.

Sikorsky delivered 174 S-70, S-76, and S-92 helicopters in 2007, double the output of just three years earlier. The company acquired PZL Mielec in 2007 and set about modernizing the 71-year-old Polish fixed-wing aircraft manufacturer to produce the S-70i, a new Black Hawk for the international market. "Sikorsky has taken a major step toward becoming a leading member of the European aerospace community," said Sikorsky President Jeffrey Pino. In 2008, Shanghai Sikorsky and Aviation Industries of China began building S-76 airframes in the People's Republic of China.

The single-seat X2 demonstrator flew first in August 2008 with a suite of technologies promising cruise speeds around 250 kt with helicopter low-speed handling and hover efficiency. In 2009, Sikorsky demonstrated automated oil rig approach technology for the big S-92. The company by that year had nearly 17,000 employees and annual revenues of $5.4 billion.

In 2010, Sikorsky Stratford delivered the 200th UH-60M Black Hawk to the U.S. Army. The X2 surpassed 250 kt in September, and Sikorsky began company-funded work on the S-97 light tactical helicopter demonstrator using X2 technologies. Tata Advanced Systems Ltd. started building S-92 cabins for Sikorsky in Hyderabad, India. The digitally-designed CH-53K also integrated major structures made by outside suppliers. Spirit AeroSystems in Wichita, Kansas rolled out the fuselage of the first CH-53K ground test vehicle. Aurora Flight Sciences delivered the first CH-53K main rotor pylon in April 2011.

By 2011, UH-60Ms were in production or under contract for Bahrain, Jordan, the UAE, the Mexican Federal Police, Sweden, the Saudi Land Forces, Taiwan, Thailand, and Brazil. Australia chose the MH-60R as its next-generation, multi-role naval combat helicopter. Sikorsky Mielec in Poland delivered its first S-70i to the Kingdom of Saudi Arabia Ministry of Interior, and Black Hawks made in Poland were on order for Mexico and Brunei. Thailand received two MH-60S Knight Hawks in 2011.

The X2 team earned the Collier Trophy in March 2011. By then, 15 S-92s were flying head-of-State missions around the world. The first CH-148 Cyclone in interim configuration arrived at Shearwater, Nova Scotia in May 2011 to train Canadian Forces aircrew and technicians.

The U.S. Army received its 500th UH-60M in 2012, but declining defense budgets led Sikorsky to close the military completions center in Horseheads and transition the work to West Palm Beach. International orders for Sikorsky helicopters grew. In December 2012, the U.S. Navy announced Denmark as the second international customer for the MH-60R. Australia took delivery of the first of 24 MH-60Rs in 2013 to replace its S-70B-2 Seahawks.

The Sikorsky-funded S-94 or X2 technology demonstrator integrated the advancing blade concept with structural, propulsion, and flight control advances.

In the commercial helicopter business, the S-76D with high-lift composite rotor blades, Pratt & Whitney Canada engines, and a Thales integrated cockpit went on the market and won orders from operators in China, Japan and Mexico. Sikorsky finished 2012 with a backlog of nearly $3 billion— its largest ever—driven by growing demand in the oil and gas industry.

The CH-53K Ground Test Vehicle at West Palm Beach beganturning its rotor head without blades attached in December 2013 and started shakedown testing with blades installed in April 2014. Combined production of UH-60M utility and HH-60M Medevac Black Hawks at Stratford for the U.S. Army was running eight to 10 aircraft per month. UH-60Ms for the Republic of China were being built at West Palm Beach. Sikorsky signed agreements with the Turkish government and local suppliers to license-produce 109 T-70 Black Hawks in Turkey. The company also teamed with Boeing to design and build the SB>1 (S-100) Defiant joint multirole technology demonstrator for the U.S. DoD Future Vertical Lift (FVL) initiative. FVL envisioned a family of fast, long-range rotorcraft, ultimately to replace the Black Hawk and other DoD helicopters.

The S-95 or Marine Corps CH-53K King Stallion designed to replace the CH-53E entered development flight test in October 2015.

The first fully-configured S-76D was delivered to the Bristow Group in January 2014. In that year, Sikorsky Coatesville delivered its peak output of 42 S-92s. Cancellation of the Presidential VXX program also positioned the S-92 as the new Marine One to replace VH-3Ds and VH-60Ns. The U.S. Navy announced on May 7, 2014 that Sikorsky would build the VH-92A fleet totaling 21 operational aircraft. Sikorsky President Mick Maurer stated, "For 57 years, our company has been trusted with the critical responsibility of building and supporting a safe and reliable helicopter fleet for the President of the United States. . . We stand ready to deliver the next Marine One, the world's most advanced executive transport helicopter." By the end of 2014, Sikorsky had over 15,000 employees and net sales of $7.5 billion.

Fast and Smart

Sikorsky Aircraft worked with the Army Avation and Missile Research Development and Engineering Center in 2014 to demonstrate autonomous flight with fly-by-wire flight controls, first on the JUH-60A Rotorcraft Aircrew Systems Concepts Airborne Laboratory (RASCAL) and later on a fly-by-wire UH-60M Upgrade (UH-60MU) Black Hawk. In March 2014, the Manned/Unmanned Resupply Aerial Lifter (MURAL) demonstration put the UH-60MU under the control of Sikorsky Matrix autonomy software and a ground control station to deliver sling loads accurately commanded by an engineer on the ground. Sikorsky meanwhile continued Matrix autonomy development on the fly-by-wire S-76 SARA -- Sikorsky Autonomy Research Aircraft and announced plans to fly an autonomous Black Hawk to mature the technology.

The S-97 Raider compound helicopter flew for the first time at West Palm Beach on May 22, 2015. The company-funded demonstrator integrated the rigid rotor advancing blade concept with fly-by-wire flight controls, integrated tail thruster, and active vibration control to achieve unprecedented speed and agile maneuvering for FVL. The last of 275 MH-60S Knight Hawks for the U.S. Navy rolled out in 2015. However, the first of four CH-53K Engineering Demonstration Models, EDM-1, hovered on October 27, 2015 at

The S-76 Sikorsky Autonomy Research Aircraft was the first company-funded testbed for autonomous flight controls.

West Palm Beach. Marine Corps plans called for 200 King Stallions. With healthy long-term prospects, Sikorsky was sold by United Technologies to Lockheed Martin Corp. for more than $9 billion.

In March 2016, Sikorsky formally opened its commercial helicopter Customer Care Center in Trumbull, Connecticut to use data from the advanced S-92 Health and Usage Monitoring System (HUMS). HUMS data establishes trends regarding component wear, maintenance and service life. Sikorsky Stratford rolled out the 1,000th UH-60M Black Hawk in October 2016, and production continued for the U.S. government, Foreign Military Sales, and direct commercial customers. MH-60Rs were being built in Stratford and integrated in Owego for the U.S. Navy and Denmark. Chile ordered six S-70i Black Hawks from the Polish production line.

Four CH-53K King Stallion engineering development model aircraft and three system development test articles were built in Florida. When Sikorsky came to terms with the State of Connecticut in 2016 to locate CH-53K production at Stratford, the company began rearranging H-60 production space to accommodate the new "Kilo." Each King Stallion build station filled the space of two Black Hawk stations.

Sikorsky and the U.S. Army finalized the ninth-Black Hawk multiyear contract in June 2017 buying helicopters for the U.S. Government and Foreign Military Sales to Taiwan, Saudi Arabia, Latvia and Thailand. The United Arab Emirates received the first armed S-70i, and commercial S-70i helicopters outfitted with firefighting water tanks became Fire Hawks.

In January 2018, Sikorsky Coatesville delivered the 300th production S-92. The Sikorsky S-97 Raider light tactical coaxial compound helicopter exceeded 200 knots in October 2018 at West Palm Beach. The bigger Sikorsky-Boeing SB>1 Defiant joint multirole technology demonstrator flew for the first time on March 21, 2019 to help the U.S. DoD make informed decisions for FVL. The U.S. Air Force HH-60W Combat Rescue Helicopter also flew for the first time on May 17, 2019 in West Palm Beach. By the end of 2020, Stratford had delivered 831 UH-60M utility and 281 HH-60M medevac helicopters to the U.S. Army. Foreign military sales in process included UH-60Ms for Saudi Arabia, Thailand, Latvia, Croatia and Jordan. The Philippines took delivery of its first Polish-built S-70i. The Republic of Korea joined the United States, Australia, Denmark, Saudi Arabia, India and Greece in the family of MH-60R Naval Hawk operators.

Sikorsky concluded the first 20 years of the 21st century with a new corporate owner, strong international partnerships, and a vibrant technology portfolio for the global helicopter market.

The S-92 Raider applied X2 technologies to a light tactical compound helicopter that exceeded 200 kt in October 2018.

Chapter 4
U.S. Government Missions

Presidential Transport
S-61/H-3

U.S. Army
S-70/H-60

U.S. Air Force
S-55/H-19

U.S. Coast Guard
S-52/H-52

U.S. Navy
S-61/H-3

U.S. Marine Corps
S-65/H-53

Sikorsky Serves the Presidents

"America must carry the torch of free life with all the immense value which this involves for individuals and nations, and even for the whole world."

The helicopter utilized by the President of the Unied States is strategic and symbolic, the flying office of America's Chief Executive at home and abroad. Sikorsky Aircraft has engineered, built, and support-ed the Presidential helicopters since 1957. The Sikorsky VH-3Ds and VH-60Ns of Marine Helicopter Squadron One (HMX-1) provide safe, timely, secure vertical lift for the Commander-in-Chief and other dignitaries as ordered by the White House Military Office. The polished "White Tops" built in Stratford, Connecticut are storied aircraft, modernized at the factory in accordance with a Special Progressive Aircraft Rework (SPAR) program. In 2017 alone, 50 years after the first official Presidential helicopter flight, they logged 6,300 flight hours in 160 Presidential lifts. The VH program marked a new chapter in June 2019 when Sikorsky received a low-rate initial production contract for the next Marine One.

Dwight Eisenhower became the first U.S. President to ride in a helicopter when he flew from the White House South Lawn to Camp David, Maryland in a three-seat Bell Model 47 (Air Force H-13J) on July 12, 1957. Operation Alert evacuated the President from Washington in a simulated nuclear attack, but a real emergency that September rushed the vacationing Eisenhower on a Sikorsky S-58 (Marine HUS-1) from Newport, Rhode Island to his fixed-wing Air Force One at Quonset Point, Rhode Island, for return to Washington, D.C. The White House soon after asked Sikorsky for an executive-appointed S-58 (Army VCH-34) built in Bridgeport, Connecticut with 12 airliner-style passenger seats, carpeting, curtains, and emergency floats.

Overall green Sikorsky S-58s equipped the Army Executive Flight Detachment and Marine Corps HMX-1 in 1958.

The 14,000 lb S-58 was powered by a single Wright R-1820-84 piston engine and later introduced automatic stabilization equipment. *Sikorsky News* quoted Presidential pilot Army Major William Howell saying, "That equipment certainly takes the strain out of flying helicopters, and its automatic features contribute considerably to flight safety."

The Army stood up its Presidential helicopter detachment in the fall of 1957 with four VCH-34s at Davison Army Airfield in Fort Belvoir, Virginia. Marine experimental squadron HMX-1 meanwhile received five similar HUS-1Zs for a Presidential Flight Detachment at Quantico, Virginia. The Army formally named the Executive Flight

Detachment in the summer of 1958 and with the Marines kept five helicopters on 24-hour alert at Anacostia Naval Station in Washington.

The Army and Marine Corps shared the Presidential mission for 19 years and with the President aboard used the radio call signs Army One or Marine One, depending on the pilot in command. Soon after the Army olive and Marine green S-58s entered service, Sikorsky technical representative Harry Asbury suggested a white-top paint scheme to help cool the Presidential cabin. To this day, HMX-1 distinguishes White Top aircraft for the President from Green Tops for support and test missions. White Tops are maintained and modernized with isolated parts in secure facilities by squadron and factory personnel with Yankee White security clearances.

In September, 1959, President Eisenhower took Soviet Premier Nikita Khrushchev on a 30-minute aerial tour of Washington DC aboard Marine One. *Sikorsky News* reported the President told the Premier, ". . . he uses helicopters every chance he gets."

President Eisenhower took Soviet Premier Khrushchev on a tour of Washington in a Marine HSS-1Z White Top in 1959.

In December 1959, a mix of Executive Flight Detachment aircraft and Fleet Marine helicopters supported Eisenhower on a 20,090-mile tour of 11 European, Asian, and North African countries in 19 days. The President arrived in foreign capi tals by Air Force One, rode to ceremonies by car, and returned to airports by helicopter to save time.

In February and March 1960, Eisenhower visited Brazil, Argentina, Chile, and Uruguay supported by 15 helicopters from HMX-1. A formation flight at 14,000 ft over the Andes Mountains forced crews to wear oxygen masks. By January 1961, Army and Marine S-58s had flown 100 Presidential missions and the two overseas tours. Army One acquired a cabin air conditioner from a yacht at the request of Mamie Eisenhower. The Kennedy Administration continued to use the piston powered VCH-34s and HUS-1Zs, re-designated VH-34C/Ds in 1962. The first Army One is preserved in the U.S. Army Aviation Museum at Fort Rucker, Alabama.

John F. Kennedy started his presidency with Marine HUS-1Zs and Army VCH-34s in 1961.

Sikorsky Sea Kings

The first Sikorsky S-61 (Navy YHSS-2) Sea King prototype powered by twin General Electric T58 turboshafts flew on March 11, 1959 and gave the Navy a new antisubmarine helicopter with record-setting performance and enhanced safety. In January, 1962, the joint Marine Corps-Army Executive Flight Detachment announced plans for eight Presidential HSS-2Zs. The 19,000 lb helicopters (soon redesignated VH-3As) were built in Stratford, Connecticut with picture windows on the left side, an auxiliary power unit on

the right sponson, and a soundproofed, air-conditioned executive interior with radio telephones, a wet bar, and toilet.

President Kennedy had his first ride in a VH-3A in May, 1962. In February 1964, four VH-3As – three Marine and one Army – flew cross-country to Los Angeles and later Palm Springs for President Johnson to meet with Mexican President Mateos and former U.S. President Eisenhower. Johnson later split the Army Executive Flight Detachment to keep at least one VH-3A near his Texas ranch.

Kennedy transitioned to the relatively spacious Army One VH-3A in 1962.

In July 1969, a VH-3A transported President Nixon 200 miles over water to the aircraft carrier USS Hornet for the recovery of the Apollo XI astronauts returning from the moon. Two more Navy SH-3As were subsequently converted to executive transports. Presidential Sea Kings had to be partly disassembled for deployment via Air Force jet transports. Teardown for "Glory Missions" took about eight hours and reassembly around 11 hours, plus a five-hour "penalty flight" for safety. In 1969, President Nixon took three VH-3As to London, Belgium, and Rome via C-5A Galaxy transport. He also took VH-3As to Cairo in June 1974 and gifted one air-conditioned helicopter to Egyptian President Sadat. (That long-serving aircraft was eventually refurbished in the U.S.and returned to Egypt in May 2009.)

The VH-3A gifted by President Nixon to Egyptian President Sadat was refurbished by Clayton International and returned to Egypt in 2009. (U.S. Navy)

Nixon rode Army One from the White House to Andrews Air Force Base after his resignation on August 9, 1974; that VH-3A remains on display at Nixon's Presidential library in California. An additional White Top Sea King was restored and is in the National Naval Aviation Museum in Pensacola, Florida.

Sikorsky News in October 1973 reported on two VH-3As returned to HMX-1 after SPAR in Stratford. These overhauls refurbish components to blueprint tolerances rather than normal repair standards, and they replace parts with finite service lives at half the normal removal times. SPAR also incorporates Presidential fleet improvements. One of the SPAR VH-3As was the first executive Sea King delivered with the bifilar vibration absorber, bigger tail pylon, and larger tail rotor introduced on commercial S-61s and later Navy SH-3s. It was a forerunner of 11 new-build VH-3Ds delivered from December 1974.

Lyndon Johnson split the Army Executive Flight Detacment to keep a VH-3A near his Texas ranch.

Sikorsky engineering remedied a VH-3D problem early in the Gerald Ford presidency. The six-foot-tall President frequently bumped his head boarding the helicopter. VH Program Manager Bill Hahn later told industry magazine *Rotor & Wing*, "Then someone on my staff came up with a simple solution: Install a rubber bumper around the top of the helicopter door. That way if President Ford banged his head against the aircraft he would be hitting it against a rubber cushion instead of the aluminum side of the helicopter."

In the spring of 1976 during the Ford Administration, a Department of Defense analysis recommended HMX-1 consolidate the Presidential helicopter fleet to save aircraft and personnel. Gerald Ford made the last flight on Army One, and the VH-3As were retired by VH-3Ds by the end of 1976. The VH-3Ds have moved Presidents and dignitaries ever since. One flew British Prime Minister Margaret Thatcher to and from Camp David for talks with President Reagan in 1986, another landed Pope Francis at the Downtown Manhattan/Wall Street Heliport in 2015 and Marine One glassware signed by President Jimmy Carter is still sold online today.

The VH-3D was upgraded with Carson composite main rotor blades. (Carson Helicopters)

The 21,500 lb VH-3D dispensed with the folding rotors of Navy sub-hunters and added a rear left cabin door with stairs. Through successive SPAR cycles, it acquired more powerful T58-GE-400B engines, high-lift composite main rotor blades, inertial and GPS navigators, the Traffic Collision Advisory System, survivability equipment, and crash-survivable flight recorders. With the President aboard, the VH-3D is the command and control platform of the U.S. Commander-in-Chief, and the Marine One communications suite includes secure and non-secure systems hardened against electromagnetic pulse.

The VH-3D retired the VH-3A in 1976 and has equipped HXM-1 ever since.

A SPAR completed in 2002 increased VH-3D airframe life from 10,000 to 14,000 flight hours. The retractable landing gear of Presidential Sea Kings is now secured in the down position to save weight and maximize availability.

White Hawks

The VH-3D with its smooth ride and stand-up cabin big enough to seat 14 passengers gave the Executive Flight Detachment a helicopter popular with Presidents. (White Top utilization traditionally spikes in election years.) The smaller Sikorsky S-70 nevertheless offered a new deployable, survivable, crashworthy aircraft well-suited to contingency missions abroad. The 22,000 lb VH-60N mixed the energy-absorbing structures and landing gear of the Army Black Hawk with the Automatic Flight Control System and folding main rotor and horizontal stabilator of the Navy Seahawk. It enhanced Presidential safety with weather radar and engine exhaust suppressors. The new White Hawk cabin seated 11 and introduced limousine-like doors with power-folding steps. Delivery of the first VH-60N was acknowledged by President Ronald Rea-

gan with a letter of thanks to Sikorsky workers in November 1988. Reagan rode the new Marine One to Camp David in January 1989.

The first VH-60N White Hawk was delivered in 1988

During Operation Desert Shield in November 1990, President George H.W. Bush landed in a VH-60N on the amphibious assault ship USS Nassau to offer Thanksgiving greetings to military personnel in the Persian Gulf. He also arrived by White Hawk in troubled Mogadishu, Somalia in December 1992. The White Hawks self-deploy to domestic emergencies and fold without disassembly to fit C-17 and C-5 transports on Phoenix Banner missions around the world. A VH-60N carried President Bill Clinton over flooded North Carolina after Hurricane Floyd in September 1999. In September, 2005, a White Hawk landed President George W. Bush aboard the LHD USS Iwo Jima for briefings on Hurricane Katrina relief efforts in and around New Orleans.

George H.W. Bush arrived by VH-60N at the American Embassy Compound in Mogadishu, Somalia in 1992. (U.S. Army)

Nine White Hawks were delivered to replace the VH-1N Hueys then with HMX-1; eight still serve the President today. A VH-60N mid-life upgrade completed in 2002 upgraded White Hawk mission equipment. Marine White Hawks now have the same General Electric T700-GE-401C engines used in Navy Seahawks. A VH-60N cockpit upgrade in 2012 gave them a digital "glass" cockpit built around the Common Avionics Architecture System used in the latest Army Special Operations Black Hawks and modernized Coast Guard Jayhawks.

VH-60Ns remain visible tools of U.S. foreign policy, deployed in detachments of two or three aircraft with 20 or 25 Marines. President Barack Obama deplaned Air Force One and boarded Marine One in Saudi Arabia in April 2016 to meet with allies about combatting the Islamic State. In November 2017, VH-60Ns shuttled President Donald Trump during crisis talks in South Korea.

George W. Bush landed in a VH-60N after Hurricane Katrina struck New Orleans in 2005. (U.S. Navy)

New 92's

President George W. Bush rode Marine One to lower Manhattan days after September 11, 2001. The magnitude of new threats led the White House to fast-track a VXX Presidential Helicopter Replacement in 2002. With aging VH-3Ds and VH-60Ns limited in performance and growing more expensive to operate, the Navy and Marine Corps sought a new helicopter with lower operating and support costs, and one able to add modern survivability equipment

and Oval Office-in-the-sky communications. The Collier Award-winning S-92 was certified by the Federal Aviation Administration (FAA) in July 2004, and Sikorsky launched a VH-92A All-American tour with a stretched S-92 demonstrator to compete for the VXX contract. The stakes were high. Sikorsky President Steve Finger stated, "The government is not simply buying Presidential helicopters off the shelf. In fact, the winner receives hundreds of millions of dollars for research and development to advance their helicopter." A Lockheed Martin - Agusta Westland bid to import, integrate, and re-engineer the Anglo-Italian EH101 in two concurrent increments won the VXX System Design and Development contract in 2005, but the three-engined VH-71A/B Kestrel was canceled in 2009 due to ballooning costs. The effort nevertheless built a new Presidential Helicopter Support Facility for future SPAR work at Naval Air Station Patuxent River, Maryland.

The Naval Air Systems Command (NAVAIR) re-evaluated VXX requirements and formulated a streamlined plan that integrates mature mission systems and a 14-passenger executive interior into the rugged, flaw-tolerant S-92A, yet preserves the helicopter's FAA airworthiness certification. Sikorsky received the VH-92A Engineering and Manufacturing Development contract on May 7, 2014. The 27,000 lb Presidential helicopter with an enhanced environmental control system is shielded against electromagnetic interference and pulse from nuclear events. It uses uprated GE CT7-8A6

Sikorsky stretched an S-92 demonstrator for the original VXX competition.

engines with added high-and-hot power for HMX-1 administrative and contingency missions.

The first VH-92A Engineering Development Model (EDM 1) first flew on July 28, 2017 at Stratford and soon after went to Sikorsky Systems Integration in Owego for instrumentation. EDM 2 followed with its first flight on November 11, 2017. Four System Demonstration Test Articles (SDTAs) built in Coatesville, modified in Stratford, and missionized in Owego are earmarked for Initial Operational Test and Evaluation and will ultimately join the Presidential fleet.

NAVAIR plans to follow the SDTA helicopters with 17 VH-92A production aircraft delivered to HMX-1 through 2023. The 21 new VH-92As will ultimately retire the historic VH-3Ds and VH-60Ns. Donald Trump first toured the new Presidential helicopter in June, 2019, when a VH-92A test aircraft practiced landings on the White House South Lawn.

On July 4, a White Top VH-92A led two Green Top MV-22s in an HMX-1 formation over the nation's capital.

A VH-92A Engineering and Manufacturing Development aircraft practiced landings on the White House South Lawn. (U.S. Dept. of Defense)

Sikorsky Lifts the Army

"For the first time anywhere, great numbers of lives were saved by helicopters during the war in Korea where thousands of lives were saved and hundreds of men rescued from behind enemy lines."

Dust-Off or Medevac Black Hawks (here, the HH-60M) have saved lives on battlefields from Grenada to Iraq. (U.S. Army)

Igor Sikorsky and pilot Les Morris delivered the experimental XR-4 (VS-316) helicopter to the U.S. Army Air Corps at Wright Field on May 17, 1942. Their flight from Bridgeport, Connecticut to Dayton, Ohio covered 761 miles in 16 hours, 10 minutes logged over five days. In *The Story of the Winged S*, Sikorsky later wrote, "It would be right to state that with the successful flight of the XR-4 in the summer of 1942, the helicopter became a reality in the United States. Its practical value and potential possibilities were proven beyond any trace of doubt."

The Army saw helicopter possibilities soon after World War I, but hover tests by the Air Service Engineering Division in 1922 concluded a 24-bladed de Bothezat quadcopter was too complicated to develop. Igor Sikorsky flew his single-main-rotor VS-300 (S-46) for the first time on September 14, 1939. In July 1940, he invited Army Air Corps helicopter project officer Major Frank Gregory to fly the one-of-a-kind demonstrator. Gregory wrote in *Anything a Horse Can Do – The Story of the Helicopter,* "After about eight minutes of flying I finally got the craft back on the ground, much to the relief of Sikorsky; although, of the group which had been watching my flight, he was most enthusiastic in his congratulations."

Igor Sikorsky, Orville Wright, and Col. Frank Gregory acknowledged delivery of the VS-316 (Army XR-4) to the Army at Wright Field, Ohio in 1942.

Gregory became the first American military helicopter pilot. The Vought-Sikorsky Division of United Aircraft Corporation proposed a two-seat VS-316 (S-47) to the Army and received a $50,000 development contract -- the Army wanted an alternative to the Platt-LePage XR-1 with its outrigger rotors. Then-Colonel Gregory accepted the experimental XR-4 on May 30, 1942. That June, Secretary of War Henry Stimson directed Army ground forces to attach fixed-wing observation aircraft to artillery units and began organic Army aviation.

Gregory's XR-4 evaluation earned Sikorsky an Air Corps contract for 15 developmental YR-4As on December 21, 1942. A production contract in May 1943 covered 100 more powerful R-4Bs. The first YR-4A was delivered on July 3, 1943. Sikorsky separated from Vought that year and began R-4 (S-47) production in a renovated and expanded Bridgeport factory. The 100th helicopter came off the line on September 7, 1944. That same month, the Army Air Forces established a helicopter school at Freeman Field, Indiana. *Sikorsky News* reported on an impromptu medical evacuation mission on October 20, 1944 when an R-4 landed near dense woods to recover an injured student pilot.

Off to Wars

The 2,500 lb R-4B was acquired by the Army as a training helicopter. Sikorsky designed the 5,500 lb R-5 (S-48) observation helicopter around a 1942 U.S. Army Air Forces (USAAF) lift requirement. The first two-seat YR-5A flew on August 18, 1943, and the Army ordered 26 helicopters for testing in March 1944. However, with the bigger R-5 in development, the 1st Air Commando Group took six underpowered YR-4s to the China-Burma-India theater over Greg-

Lt. Carter Harman flew the S-47 (Army YR-4B) in the first combat rescue in 1944.

ory's objections. Lt. Carter Harman flew the first combat rescue, recovering three wounded British soldiers and their American pilot down behind enemy lines with four sorties over two days in April 1944.

The 2,600 lb R-6 (S-49) flew on October 15, 1943 and went to war alongside R-4s shuttling bomber parts to and from the Army's Ivory Soap depot ships in the Pacific. On June 15, 1945 the 5th Aircraft Repair Unit launched an R-4 to evacuate two wounded soldiers. From June to July 1945, R-4s and R-6s evacuated 75 to 80 wounded from the difficult highlands northeast of Manila.

The USAAF awarded Sikorsky a production contract in 1944 for 100 R-5D rescue helicopters. With the Bridgeport factory committed to the R-5, the USAAF Production Division ordered manufacture of the smaller R-6 licensed to Nash-Kelvinator in Detroit, Michigan. The end of World War II in August 1945 canceled major aircraft orders. Only 193 production R-6As were made for the USAAF, and some were sold to the U.S. Navy and U.K. Royal Air Force.

The S-49 (Army R-6) flew medical evacuation missions in the Philippines at the end of World War II.

The USAAF ultimately received 14 R-5A and 20 R-5D helicopters from July 1945 to October 1946. Aviation units fell under the Army Transportation Corps in 1946. However, when the U.S. Air Force (USAF) became an independent service in 1947, new Army aircraft had to be acquired through USAF channels. The struggle over roles and missions led the Air Force to refuse an

Army request for helicopters in 1948. Sikorsky first flew the 6,700 lb YH-19 (S-55) in November 1949, but the Air Force blocked Army orders in 1950. While the Marines deployed S-55s to mountainous Korea in April 1951, the Army's 6th Transportation Company (Helicopter) was not activated with H-19 Chickasaws until July 1952.

With room for up to eight troops or six casualty litters, H-19s were the Army's first transport helicopters. They flew their first Korean mission on March 20, 1953, when the 6th Transportation Company hauled 17 tons of supplies to elements of the 3rd Infantry Division. The 13th Transportation Company joined the 6th, and in late June the two units formed an air bridge to supply isolated infantry. More actions followed, and before the Armistice, Army H-19s flew in Operation Little Switch to repatriate sick and wounded American POWs.

Back in the U.S., Exercise Snowstorm in February and March 1953 took 11 H-19s of the 506th Transportation Company from Fort Benning, Georgia to Camp Drum, New York, the first time Army cargo helicopters flew in an extended field exercise. In June 1954, the 328th Transportation Helicopter Company with H-19s enabled the

The S-55 (Army H-19 Chickasaw) gave the Army an interim cargo helicopter for the Korean War and the formative 1950s.

Seventh Army in Europe to experiment with air mobility tactics and techniques. In 1957, Colonel Jay Vanderpool, Chief of the Army Aviation Schools combat development office, built a Sky Cav platoon at Fort Rucker, Alabama including an armed H-19. The Army ultimately received 72 H-19Cs and 240 H-19Ds that served into the 1960s.

The S-58 (Army H-34 Choctaw) was the "light" cargo helicopter used to develop early airmobile concepts.

Heavy Haulers

Army Aviation plans in the early 1950s classed the H-19 (S-55) as an interim cargo helicopter. The Army followed the Marine Corps lead for its "light" H-34 (Sikorsky S-58) and "medium" H-37 (Sikorsky S-56) cargo helicopters. A 46,700 lb tandem-rotor "heavy" cargo helicopter was canceled before production. The Army ordered the 14,000 lb S-58 in 1953, borrowed Marine helicopters for testing, and took delivery of its first H-34 Choctaw in April 1955. Unit helicopter training on H-19s and H-34s moved from Fort Sill, Oklahoma to Camp Rucker, Alabama. The Army Aviation Center was established at Fort Rucker in 1955.

The Choctaw was soon deployed with U.S. Army-Europe, and in January 1958, 47 H–34s moved 950 soldiers and 15 tons of equipment from Heil-bronn, Germany to the Baumholder training area in Operation Lion Lift. *Sikorsky News* quoted Eighth Infantry commander Maj. Gen. P.F. Linde-man, "It is a good example of what Army Aviation can do under difficult conditions. It proved, too, that a tactical combat lift can take place even under worse conditions than we had anticipated." In a joint Executive Flight Detachment, Army and Marine H-34s became the first helicopters to transport the President of the United States routinely. The Army received 437 Choctaws through 1965 and returned early H-34s to Sikorsky for automatic stabilization equipment.

The "medium" S-56 (Marine XHR2S) first flew at Bridgeport in December 1953. The Army borrowed one YH-37 from the Marine Corps for testing in 1954 and took delivery of its first CH-37A Mojave in December 1956 at Fort Rucker. The twin-engine, 31,000 lb cargo helicopter could load 26 combat troops, 24 litter patients or up to 10,000 pounds of cargo through clamshell nose doors. The H-37 could also airlift a Little John rocket, launch crew, and tow jeep internally – the Armair Brigade proposal in 1956 used the nuclear-capable Little John for long-range aerial artillery support.

In February, 1958, the 4th Transportation Company (Medium Helicopter) at Fort Benning, Georgia became the first unit equipped with the H-37.

The twin-engine S-56 (Army H-37 Mojave) was a "medium" cargo helicopter with cabin space for 26 troops and power to carry heavy cargo.

Sikorsky News reported in July 1958 on Project AMMO at Fort Bliss, Texas and White Sands, New Mexico where three H-37As sling-lifted a nuclear-capable Honest John rocket, launcher, support trailer and prime mover to a launch demonstration. In the same exercise, the H-19s and H-34s of a notional Sky Cav fired guns and rockets.

The 4th Transportation Company deployed the H-37 to Germany in 1959 and the 90th Transportation Company in 1961. Sikorsky concluded S-56 production in May 1960, but 90 of 94 H-37As returned to Stratford through 1962 to become H-37Bs with auto-stabilization equipment. Between 1963 and 1966, nine Army CH-37Bs served in Vietnam sling-lifting downed aircraft. The big, piston-engined Mojave retired from Army National Guard service in 1969, a placeholder pending more powerful turbine-engined helicopters.

Modified from the S-52 (Army XH-18), the fast S-59 (Army XH-39) was Sikorsky's first turbine-engined helicopter.

Sikorsky's first turbine-engined helicopter -- the experimental four-seat XH-39 (S-59) with a 400 shp Continental/Turbomeca Artouste turboshaft -- set a world helicopter speed record at 135.6 kt in August 1954 and an altitude record at 24,500 ft later that year. It nevertheless lost the pivotal competition for a turbine-powered air ambulance that would later fill Army airmobile units.

The S-60 Skycrane flew in 1959 with twin Pratt & Whitney R-2800 radial engines on a simple airframe with no cargo cabin. It demonstrated the lifting potential of the crane helicopter all through 1960. United Aircraft approved turbine-engined S-64 development in April 1961. Sikorsky flew the S-64 with twin Pratt & Whitney JFTD-12A-1 turboshafts on May 9, 1962 and in July delivered the company-funded demonstrator to Fort Benning, Georgia for Army evaluation.

The Tactical Mobility Requirements Board headed by General Hamilton Howze in 1962 formulated new helicopter air mobility concepts. The Army ordered six YCH-54s for testing and accepted the first Tarhe at Stratford, on June 30, 1963. The 478th Flying Crane Company, 44th Air Transportation Battalion deployed the big helicopters to Vietnam in 1965 and supported the 1st Cavalry Division (Airmobile) moving bulldozers and other heavy equipment and recovering downed aircraft. In Operation Masher-White Wing, CH-54s first airlifted 155 mm howitzers to firing positions. Skycranes in Vietnam hauled mobile command posts and dropped 10,000 lb bombs to clear jungle landing zones.

The company-funded S-60 demonstrated the potential of the crane helicopter, here carrying an Army H-34.

In April 1965, a CH-54A from the 478th Aviation Company carried a "people pod" packed with 87

The S-64 (Army CH-54 Tarhe) gave the Army a Skycrane with turboshaft power to lift heavy external loads.

troops. On December 30, 1968, Army pilots flew a CH-54A from Sikorsky's Stratford plant through 30,000 ft to break helicopter altitude records. The CH-54A was succeeded by the more powerful CH-54B in 1969, and the big Sikorsky cranes remained in service until the last was retired by the Nevada Army National Guard in January 1993.

Hardened Helicopters

Lessons learned from helicopter combat in Southeast Asia helped define the crashworthy, ballistically tolerant Utility Tactical Transport Aircraft System (UTTAS) specified by the U.S. Army in 1971. Sikorsky flew the YUH-60 (S-70) with twin General Electric T700 turboshafts in October 1974, and won the competition in December 1976. In his book, *Black Hawk – the Story of a World Class Helicopter,* UTTAS program engineering manager Ray Leoni observed, "That award was the beginning of a new era in Army air mobility, and it brought about a major turnaround in Sikorsky's prospects for the future." Since the UTTAS decision, more than 5,000 S-70 helicopters have fought wars and saved lives with the U.S. Army, Navy, and Air Force, and Coast Guard, allied air arms, and civil agencies.

The Army received the first production UH-60A

The S-70 (Army UH-60) Black Hawk won the UTTAS competition and gave the Army an evolving helicopter for air assault, Medevac, and Special Operations.

Black Hawk in October 1978 and took the 16,800 lb, twin-turbine squad carrier to war in the hurried invasion of Grenada in October 1983. During the Caribbean assault, the heavy-duty control components, redundant control runs, and self-sealing fuel tanks of the Black Hawk proved their value. One UH-60A with a wounded pilot kept flying with 45 bullet holes in the airframe, two in the main rotor, and one in the tail rotor.

Operation Urgent Fury also marked the special operations debut of the UH-60A with the 160th Aviation Battalion. In 1986, Special Operations pilots picked up new Black Hawks at Stratford for conversion to MH-60As and began a night/adverse weather evolution that continues today with the MH-60M in the 160th Special Operations Aviation Regiment (Airborne).

The Army UH-60L entered production in October 1989 and demonstrated it could reposition a 105 mm howitzer, six gun crew, and up to 30 rounds of ammunition in a single lift. Black Hawks again went to war in Panama during Operation Just Cause in December 1989. Stripped of crashworthy troop seats, they sometimes flew 25 soldiers into battle at a time.

Sikorsky had delivered more than 1,000 Army Black Hawks by 1990 when UH-60s were rushed to Operation Desert Shield in Southwest Asia. When Desert Shield became Desert Storm in January 1991, Black Hawks hauled troops, equipment, and supplies throughout the theater of war. They carried Special Operations Forces deep into Iraq, evacuated the wounded and the captured, deployed artillery, and rescued downed aircrew. The 101st Airborne Division launched a two-step air assault including 66 Black Hawks flying soldiers and equipment to landing zones deep inside Iraq to block enemy reinforcements and supplies.

In Army operations other than war, UH-60s performed with equal distinction. Black Hawks were part of the relief forces in Operation Provide Comfort after the first Iraq war, and they went on to support relief and peacekeeping operations in Somalia during Operation Restore Hope in 1992 and 1993. When floodwaters threatened lives in Texas in 1994, one Army National Guard Black Hawk evacuated more than 200 people from a flooded sub-division. When brushfires charred vast areas of Idaho and other states in the summer of 1994, one UH-60 crew set a record by dumping 102 water buckets in one day on a strategic ridge, making quick turns at high altitude and high temperatures to fill fire buckets from mountain lakes. Fifty Black Hawks were part of the initial peacekeeping contingent when the U.S. Army launched Operation Joint Endeavor in Bosnia in 1995.

An Army Utility Helicopter Fleet Modernization Analysis in 1999 led to a Black Hawk operational requirements document in 2001 calling for better performance, lower operating and support costs, new digital interoperability, more lift, and greater range. After America was attacked on September 11, 2001, Black Hawks provided essential air mobility and Dust-Off medical evacuation in rugged Afghanistan. When Army aviation returned to Iraq in 2003, Black Hawks again moved troops and evacuated wounded.

Two-front war tripled UH-60 utilization compared with peacetime norms and added urgency to UH-60M modernization plans. The Army Aviation and Missile Command awarded Sikorsky a contract in 2001 to remanufacture one UH-60A

and one UH-60L to UH-60M standards, convert another UH-60A into a MEDEVAC HH-60M, and build a new M-model Black Hawk from scratch. Remanufacturing plans soon gave way to all-new production. Sikorsky's 10th H-60 multi-year production contract will stretch through June 2027. Army aviation plans call for 956 UH-60M, 419 HH-60M, and 760 digital-cockpit UH-60V Black Hawks in 2035 when Future Vertical Lift aircraft bring transformational speed and range to the Army's Objective Force.

The S-100 (SB>1) Defiant models the Army's Future Long Range Assault Aircraft that will partially replace the Black Hawk.

The stealthy, computer-smart RAH-66 Comanche was cancelled in 2004. Today, Sikorsky is competing to build the Future Attack-Reconnaissance Aircraft (FARA) to fly the Comanche mission and the Future Long-Range Assault Aircraft (FLRAA) to succeed the Black Hawk. The proposed S-102 Raider-X FARA and S-103 Defiant-X FLRAA are based on the company's high-speed X2 technologies and draw on the heritage of Sikorsky Aircraft in the U.S Army.

The air-refuelable MH-60K Special Operations Aircraft gave the Army a long-range, night/adverse weather Black Hawk for Special Forces. (U.S. Army)

The Raider-X competitive prototype is Sikorsky's solution for a Future Attack Reconnaissance Aircraft.

Sikorsky Serves the Coast Guard

"Stories of helicopter rescues would fill volumes, and the number of lives saved is steadily increasing, I, personally, would lke to express my deepest respect and admiration for the gallant pilots and helicopter crews who perfrom these flights. Their actions, representing considerable skill and courage, equal the most heriic of battelfield achievements.
It would be right to say that the helicopter's role in saving lives represents one of the most glorious pages in the history of human flight."

Sikorsky Aircraft has been central to the 75 years of Coast Guard helicopter history.
(Igor I. Sikorsky Historical Archives)

A flight demonstration of Igor Sikorsky's VS-300A helicopter at Bridgeport, Connecticut in April 1942 started an air-sea rescue revolution in the U.S. Coast Guard. The chief of the Coast Guard Aviation Engineering Division, Commander William Kossler, and the commanding officer of Coast Guard Air Station Brooklyn, New York, Commander Watson Burton, both saw life-saving potential in the compact rotorcraft that could take off and land vertically and hover. Their vision led to successive Sikorsky helicopters of growing power and sophistication. The VS-300 lineage lives on in the digitized MH-60T Medium Range Recovery helicopters in today's busy fleet, and it may shape the next generation of Coast Guard vertical lift.

Coast Guard officers, impressed by the 1,300 lb VS-300, quickly saw multiple applications for the helicopter. Burton wanted an air-sea rescue platform and an alternative to harbor patrol blimps. Kossler's friend and early Coast Guard aviator Capt. Frank Erickson visited Bridgeport in June 1942 and advised Coast Guard commandant Adm. Russell Waesche that helicopters flying from ships could use radar and dipping sonar to hunt German submarines attacking Atlantic convoys.

Igor Sikorsky met with Coast Guard helicopter pioneer Captain Frank Erickson. (Igor I. Sikorsky Historical Archives)

Kossler held a place on the interagency board for Army-Navy/Coast Guard helicopter procurement and noted the 2,500 lb Sikorsky S-47 (R-4) in production for the Army could carry two crew, a depth charge, and fuel for four hours' sub-hunting. Coast Guard Commandant Vice Admiral Russell Waesche approved the idea, but Coast Guard aviation acquisition fell to the Navy. After meeting with Waesche in February 1943, Chief of Naval Operations Ernest King directed the Navy Bureau of Aeronautics to test helicopters on merchant ships. Waesche meanwhile ordered Kossler to build a training program for helicopter pilots and maintainers and a plan for helicopters at Coast Guard Air Stations.

Erickson trained on the S-47 (R-4) at Bridgeport and became the first Coast Guard helicopter pilot. He took delivery of the first HNS-1

(the Navy R-4) in October 1943. A month later, Erickson stood up his joint helicopter training base at Coast Guard Air Station Brooklyn, Floyd Bennett Field, with three HNS-1s to train pilots of the U.S. Army, Navy, and Coast Guard, and the British Helicopter Service Trials Unit.

The Coast Guard S-47 (HNS-1) at Coast Guard Air Station Brooklyn was used to develop helicopter rescue hoists in 1944. (Igor I. Sikorsky Historical Archives)

Igor Sikorsky's son Sergei served under Erickson at Floyd Bennett Field and recently recalled, "He saw the role that the helicopter would play as the 'flying lifeboat' for the Coast Guard. It was Erickson who began developing the helicopter rescue hoist, the rescue basket, now called the Erickson Basket, and much more." Erickson was the impassioned proponent of Coast Guard rotary-wing aviation. "His vision of the helicopter did not sit well with Coast Guard headquarters where the focus was on the fixed-wing flying boats. He was the Coast Guard's Billy Mitchell."

In open-sea trials in early 1943, British and American pilots found flying the S-47 (YR-4B) from merchant ships hazardous. In January 1944, Coast Guard helicopter pilot No. 2, Lt jg. Stewart Graham flew from a merchant ship in a North Atlantic convoy, but with early helicopters not ready for sub-hunting, Coast Guard efforts refocused on air-sea rescue.

On Aug 14, 1944, Igor Sikorsky greets son Sergei, AM-M3c, during a visit to Coast Guard Air Station, Brooklyn, N.Y. during helicopter tests.
(Igor I. Sikorsky Historical Archives)

On January 3, 1944, Erickson demonstrated the value of the helicopter when the destroyer USS Turner blew up in Ambrose Channel off Sandy Hook, New Jersey. The pioneer pilot flew an S-47 (HNS-1) through rain, sleet, and snow from South Ferry, Manhattan to a New Jersey shore hospital to deliver two cases of blood plasma in just 14 minutes. Surface travel would have taken hours. Kossler recommended Erickson for the Distinguished Flying Cross without result. However, *Sikorsky News* in November 1944 reported a plant visit by ranking Army, Navy and Coast Guard officers including Cdr. F.E. Erickson.

Hover and Hoist

In early 1944, an Army YR-4B recovered four airplane crash survivors with four landings in Japanese-occupied Burma. The Coast Guard took notice, and according to Sergei Sikorsky, "While reading the mission report, Erickson realized that a hoist-equipped helicopter could have lifted the survivors to safety far more quickly, and the concept of the helicopter rescue hoist was born." CGAS Brooklyn experimented with electric and hydraulic hoists. Igor Sikorsky

An S-47 (HNS-1) from Brooklyn flew hoist rescue demonstrations in Jamaica Bay.
(Igor I. Sikorsky Historical Archives)

himself visited the air station in August, 1944 and rode the hoist under Erickson's hovering helicopter.

The end of World War II shut the Brooklyn helicopter schoolhouse, but rescue experiments continued with the Coast Guard Rotary Wing Development Project Unit at Elizabeth City (E-City) North Carolina. On September 22 and 23, 1946, Coast Guard pilots got a real call to action. Cdr. Frank Erickson, Lt. "Stew" Graham, Lt. Gus Kleisch, and Lt. Walt Bolton were airlifted by Army C-54 transport from E-City to Newfoundland with an HNS-1 to rescue survivors of a Sabena airliner crashed 20 miles from Gander. They were joined by a new HOS-1 (the Sikorsky S-49/R-6) from CGAS Brooklyn. The two helicopters shuttled from makeshift helipads to the crash site to move 18 injured passengers to a lake for evacuation by fixed-wing amphibians.

Like Father, Like Son, just hanging out

On Aug. 14, 1944, an HNS-1 carries Igor Sikorsky-on the hoist with pilot Cdr. Frank Erickson at the controls. (Igor I. Sikorsky Historical Archives)

Summer, 1944 - Frst public demonstration of a helcopter rescue hoist. The pilot, Commander Frank Erickson "rescued" AMM 2/c Sergei Sikorsky. (Igor I. Sikorsky Historical Archives)

An S-47 (HNS-1) was airlifted to Newfoundland to rescue survivors of a remote airliner crash in 1946.
(Igor I. Sikorsky Historical Archives)

A Coast Guard S-49 (HOS-1), the Gander Express, was also deployed to Newfoundland in 1946. (National Naval Aviation Museum)

The Coast Guard subsequently ordered nine Sikorsky S-51s (HO3S-1Gs) delivered from 1946 to 1950. The 5,500 lb helicopter with a 450 hp Pratt and Whitney Wasp Jr. engine was big enough for a pilot and three passengers. Cdr. Erickson remained the chief helicopter development engineer at E-City. *Sikorsky news* reported in March 1949 that he flew an HO3S-1G for an hour hands-off thanks to stabilizing airfoils under the main rotor. E-City also tested inflatable floats for water landings. Coast Guard helicopters responded to real emergencies. In February 1950, Lt. Fletcher Brown piloted an HO3S-1G from E-City to Arkansas to taxi doctors and nurses to families cut off by St. Francis River flooding.

The S-51 (HO3S-1G) was used for rescue helicopter development at Elizabeth City, North Carolina. (Igor I. Sikorsky Historical Archives)

The small helicopters initially operated by the Coast Guard were clearly limited in performance and payload. The first of eight HO5S-1G (Sikorsky S-52-3) helicopters was delivered to the Coast Guard on September 24, 1952. The 2,700 lb helicopters proved too small and slow. All were placed in storage beginning in April 1954.

The S-52 (H05S-1G) in the Coast Guard was the first Sikorsky helicopter with all-metal rotor blades but had only a short operational career

The U.S. Air Force had sent prototype Sikorsky S-55s to Korea in March 1951, and production models of the 10-seat, 7,200 lb helicopter were ultimately adopted by all the U.S. armed services. The Coast Guard received the first of seven HO4S-2Gs with 550 hp Wright radial engines in November 1951. On January 19, 1952, LCdr. Gordon MacLane flew a new HO4S-2 from CGAS Port Angeles, Washington to rescue five survi-

vors of an Air Force SB-17 from a 5,000 ft high crash site on Tyler Peak in Washington state. Starting in January 1952, 23 HO4S-3G helicopters (later designated HH-19Gs) introduced 700 hp engines and night/instrument flight capability. Then-Cdr. Stew Graham used one for the first recorded night hoist rescue in the Gulf of Mexico in January 1955. Eight Marine Corps HRS-3s were subsequently transferred to the Coast Guard.

The S-55 (HO4S-1) gave the Coast Guard a helicopter with night/instrument flight capability. (Igor I. Sikorsky Historical Archives)

Coast Guard air-sea rescue doctrine still centered on fixed-wing amphibians, but one HO4S-3G reshaped public helicopter perceptions and overturned Coast Guard aviation plans when the Feather River flooded Yuba City, California on December 23, 1955. In 29 continuous operating hours, the Sikorsky helicopter from air station San Francisco hoisted 138 flood victims to safety. The first 58 rescues were at night. Lieutenant Henry Pfeiffer maneuvered his aircraft among trees, power and telephone lines, and TV antennas to hoist people from flooded homes and take them to high ground. Lt. Cdr. George Thometz relieved Pfeiffer at the controls without shutting the aircraft down. On one sortie, he shuttled 14 adults and children from a roof despite hovering close to a dangerous obstruction.

Nationwide media attention helped make the helicopter the dominant air-sea rescue platform of the Coast Guard. The aviation plan of 1957 included 79 medium range helicopters big enough to rescue six survivors 300 nm offshore. Sikorsky developed the 14,000 lb Sikorsky S-58 (HSS-1) sub-hunter for the Navy with a 1,525 hp Wright radial engine and automatic hover capability. The Coast Guard ordered six derivative HUS-1G helicopters in 1959.

The S-58 (HUS-1G) introduced hands-off night hover capability but was used only briefly by the Coast Guard. (Igor I. Sikorsky Historical Archives)

Turbines to the Rescue

Despite its size and power, the HUS-1G was limited by its heavy reciprocating engine, especially at high ambient temperatures. It was soon retired. Turboshafts promised more power in a far lighter package with greater reliability. Sikorsky flew twin General Electric T58 turboshafts on an HSS-1F test helicopter in February 1957. Late that same year, the company announced development of the 8,300 lb, 10-pasenger S-62 with a single T58, HO4S dynamics, and a boat hull for water landings.

The commercial S-62A became the Coast Guard HU2S-1G or HH-52A (S-62C) Seaguard with hydraulic rescue hoist and fold-down platform for water rescues. The HH-52A stability augmentation system also provided a beep-to-a-hover function that could bring the helicopter to a stable hover over water at night. The S-62A first flew in May 1958, and the Coast Guard HH-52 entered service at Air Station Salem,

The turbine-powered S-62 or HH-52A Seaguard had a boat hull and fold-down recovery platform for water rescues. (Igor I. Sikorsky Historical Archives)

Massachusetts in December 1962. The last of 99 HH-52As was delivered in 1969.

During their time on alert, HH-52As were credited with more than 15,000 "saves." HH-52As deployed during Hurricane Betsy on the U.S. Gulf Coast in September 1965 alone made 1,200 rescues. On the night of December 21,1968 LCdr George Garbe flew an HH-52A to pull five sailors from a fishing vessel aground and breaking up off Marmot Island, Alaska. Unable to climb higher because of freezing conditions, Garbe penetrated heavy snow showers and fog and executed a beep-to-a-hover approach with aircraft lights off to prevent reflections from sea spray and snow. He landed in the water about a mile from the vessel, taxied the helicopter towards the vessel until rocks appeared, and lifted off again to air-taxi over the ship and hover over the stern while avoiding surrounding terrain and ship's rigging. He returned five times to rescue the entire crew.

Sikorsky HH-52s served the Coast Guard well from 1963 to 1986, but they were limited to a rescue radius of about 150 nm and on hot days could typically carry only three survivors and crew. The Coast Guard conducted a Medium Range Recovery (MRR) helicopter competition won by Sikorsky's twin-turbine, long-body S-61R amphibian in production for the Air Force as the HH-3E combat Search And Rescue helicopter. The Coast Guard HH-3F first flew on October 11, 1967. Forty "Pelicans" were delivered from

The S-62 (HH-52A)s deployed during Hurricane Betsy in 1965 were credited with 1,200 lives saved. (Igor I. Sikorsky Historical Archives)

The big S-61 (HH-3F) gave the Coast Guard capability for long-range rescues and counter-drug operations. (Igor I. Sikorsky Historical Archives)

December 1968 to June 1972. (In 1990, the Coast Guard added nine Air Force HH-3Es and CH-3Es.)

The HH-3F First Implementation Station was CGAS New Orleans. At around 19,000 lb normal mission weight, the HH-3F with 4,000 lb normal fuel had 3 to 3.5 hours endurance plus reserves. Maximum gross weight with up to 6,000 lb internal auxiliary fuel was 22,050 lb. On March 1, 1977, Lt. James Stiles rescued four crewmen from a fishing vessel sinking off Cape Sarichef, Alaska. From Air Station Kodiak, he flew the HH-3F 475 miles under ceilings as low as 100-feet with half-mile visibility in heavy snow showers, icing, and winds gusting to 70 knots. On scene, Stiles hovered over the vessel while his crew hoisted the four sailors aboard to return in treacherous weather.

Before commercial helicopter Emergency Medical Services were widespread, long-range HH-3Fs were called upon for MAST — Military Assistance to Safety and Traffic — missions, flying emergency cases with medical teams to hospitals. The HH-3F supported ATON — Aids TO Navigation — airlifting work crews and equipment to automated lighthouses, beacons,

and buoys. The big helicopter could also carry the 4,200 lb Air Deployable Antipollution Transfer System to pump out oil tankers run aground.

The twin-turbine HH-3F gave the Coast Guard a helicopter with greater range and capability than the single-engine HH-52A. (Igor I Sikorsky Historical Archives)

The mid-1980s brought a change in Coast Guard rescue helicopter doctrine with the addition of qualified rescue swimmers to HH-3F and HH-52A crews. Swimmers were trained and equipped to deploy in storm-tossed seas and help survivors into the helicopter rescue basket. Their presence gave helicopter pilots more rescue options, especially when approaching ships with whipping masts and other hazards.

On 10 December 1987, an HH-3F from Air Station Sitka, Alaska, flew through heavy snow to a vessel taking water. The sailor and his young son abandoned their boat in survival suits. After several unsuccessful attempts to get the survivors into the rescue basket, Petty Officer Jeffery Tunks jumped into 25 to 30 ft seas. The helicopter fought 35 to 70 kt winds to recover the three from the violent sea. Tunks became the first rescue swimmer to earn the Distinguished Flying Cross.

Though the Coast Guard started deploying HH-52As on cutters in 1973 for drug interdiction, HH-3F structures were never designed for shipboard operations. However, in the 1980s, the land-based HH-3F carried drug enforcement agents for OPBAT - Operations Bahamas, Turks, and Caicos. Coast Guard crews flying from forward operating locations also inserted, extracted, and resupplied enforcement teams around Central America. CGAS Clearwater, Florida, close to the action, retired the last HH-3Fs in May 1994.

Jayhawk Generations

Sikorsky chose not to bid the 10,500 lb S-76 in the Coast Guard Short Range Recovery helicopter competition in 1979. However, Coast Guard plans for a new Medium Range Recovery helicopter coincided with a Navy requirement for a new Strike Rescue and Special Warfare Support Helicopter. Sikorsky and the Navy developed the HH-60H (S-70B-5) from the sub-hunting Seahawk, and Coast Guard officials ordered the derivative HH-60J on September 29, 1986. *Sikorsky News* in March 1989 showed the first Jayhawk in final assembly at Stratford. The 22,000 lb HH-60J with weather radar and other Coast Guard equipment was delivered in March 1990. A service-wide contest named it Jayhawk.

The Jayhawk has up to seven hours endurance

The S-70 (HH-60J) Jayhawk became the Coast Guard Medium Range Recovery helicopter in 1990. (Igor I. Sikorsky Historical Archives)

with external fuel. On the night of 28 October,1991, an Elizabeth City HH-60J crew commanded by Lt. Paul Lange flew into Hurricane Grace on a distress call from the sailing vessel Anne Kristina about 300 nm east of Cape Henry, Virginia. The Jayhawk crew refueled on the aircraft carrier USS America conducting sea trials 100 nm offshore and arrived on the rescue scene to execute an automatic precision approach to coupled hover. They located the sinking schooner with night vision goggles, dropped a rescue swimmer in 40 ft seas, and hovered in 60 kt winds and driving rain to hoist nine sailors from the Atlantic.

Sikorsky delivered the last of 42 HH-60Js to the Coast Guard in 1996. Aging airframes, obsolescent avionics, and the armed Airborne Use of Force mission after 9/11 led the Aviation Logistics Center at E-City to rebuild the Jayhawk fleet to MH-60T standards with digital "glass" cockpits, night visionics, and other improvements.

On October 29, 2012, during Superstorm Sandy, an MH-60T piloted by LCdr. Steven Cerveny rescued sailors from the sinking HMS Bounty about 90 miles southeast of North Carolina's Outer Banks. The Coast Guard crew flew in darkness, 60 knot winds, and driving rain to execute an instrument descent to the debris field of the ship. The rescue swimmer jumped into 30-ft seas to hoist one survivor. The helicopter crew found the remaining survivors in two life rafts and rescued four more sailors in high winds. Forced to withdraw with fuel low, they called in another MH-60T to recover the remaining nine survivors.

Coast Guard Commandant Admiral Karl Schultz recently notified Sikorsky President Dan Schultz that the Jayhawk fleet had logged its 10,000th save in September 2019 when MH-60Ts responded to Hurricane Dorian. Admiral Schultz concluded, "My guiding prinicples are 'Ready, Relevant, Responsive,' and the H-60 Jayhawk has helped the Coast Guard men and women uphold these tenets for nearly 30 years. Thank you for your dedication to our Coast Guard and our Nation."

Jayhawk Saves by Coast Guard Air Station

Clearwater, Florida - 3,165
Elizabeth City, North Carolina - 1,349
Sitka, Alaska - 1,165
Kodiak, Alaska - 1,093
Cape Cod, Massachusetts - 1,049
Aviation Training Center Mobile, Alabama - 983
San Diego, California - 461
Columbia River, Oregon - 432
Houston, Texas - 338
Traverse City, Michigan – 29

Jayhawks are in high demand and recently replaced Short Range Recovery Helicopters at Traverse City, Michigan. A Jayhawk crew from Traverse City rescued four sailors from a disabled boat on Lake Huron in October. The MH-60T flew about 100 miles from its home Air Station to locate the vessel in poor visibility off Rogers City. It deployed a rescue swimmer to hoist the crew amid 35 mph winds and 15 ft waves, a rescue that took more than an hour on-scene.

An MH-60T Service Life Extension Program aims to keep the Medium Range Recovery fleet operational through the mid-2030s to coincide with the Department of Defense Future Vertical Lift initiative. The Coast Guard is studying whether to replace hard-flown MH-60Ts by converting low-time Seahawks (Sikorsky S-70B-4s) to Jayhawks or to keep the helicopters flying with parts from retired Navy aircraft.

MH-60T modernization rebuilt the Jayhawk with digital avionics and night vision electro-optics. (DoD)

Coast Guard 6043 was the first MH-60T with parts of Coast Guard Jayhawk and Navy Seahawk helicopters. The so-called Frankenhawk joined the fleet in 2009. (USCG)

The Coast Guard calculates its 45 Jayhawks today have the highest average flight hours of any H-60 fleet in the world. Since 2005, the Aviation Logistics Center at E-City has converted six retired Navy SH-60F Seahawks into MH-60T Jayhawks to replenish and grow the Coast Guard fleet. HH-60J to MH-60T conversions concluded in August 2016, but demand for Jayhawk capabilities is high. MH-60Ts from Air Station Clearwater, Florida deployed to the Bahamas in September 2019 for search and rescue and relief missions after Hurricane Dorian.

By August 2018, the U.S. Navy had transferred 55 SH-60Fs to the Coast Guard. The service plans to compare the cost of additional conversions to the value of a Service Life Extension Program (SLEP) harvesting SH-60F structures. The notional SLEP aims to match remaining MH-60T service life with production of a marinized Future Vertical Lift (FVL) aircraft to replace both Medium and Short Range Recovery helicopters.

The ultimate shape of FVL Coast Guard vertical lift is to be determined. Sikorsky is flying X2 technologies today for a new generation of fast compound helicopters applicable to all the U.S. armed services, including Coast Guard life savers.

Igor Sikorsky himself introduced the cherished Winged-S award for all aviators who saved lives with their Sikorsky aircraft.

Jayhawk performance helps technicians maintain Aids To Navigation. (USCG)

Sikorsky's Air Force Lifesavers

It would be right to say that the helicopter's role in saving lives represents one of the most glorious pages in the history of human flight"

The U.S. Air Force became an independent Service in 1947 with Sikorsky S-48 (H-5D) helicopters from the Army Air Forces.

The first helicopter rescue behind enemy lines was made in Burma by the U.S. Army Air Corps in a Sikorsky S-47 (R-4) on April 26, 1944. When the Air Force gained independence in September 1947, the subordinate Air Rescue Service had about 20 S-48s (H-5Ds) built in Bridgeport for the Army Air Forces. The new U.S. Air Force (USAF) bought the civil-certified S-51 (H-5F) and sponsored further rescue developments. In 1949, an H-5G set an unofficial altitude record for an operational helicopter with a hovering rescue at 13,500 feet in California's Sierra Nevada mountains.

The August 12, 1949 *Sikorsky News* pictured the H-5H with metal rotor blades, floats, and a bulged cabin big enough for three litter patients. Late that year, the 2nd Air Rescue Squadron (ARS) on Guam received two H-5Hs -- one with pontoons for water pickups and the other with a hoist for inland rescues. The February 10, 1950 *Sikorsky News* reported, "The H-5H is currently being used by the Air Rescue Service in all parts of the globe. It was designed and built by Sikorsky Aircraft especially for use in rescue work."

At the start of the Korean War in June, 1950, the 3rd ARS had a mix of fixed-wing aircraft and nine S-51 helicopters in Japan. The squadron first deployed H-5Fs to Taegu, Korea to rescue downed aircrew and evacuate wounded soldiers – Air Force S-51s could carry two external litter capsules. In February, 1951, Lt. Daniel Miller earned the Air Force humanitarian Cheney Award for landing three times under fire and in deep snow to rescue six wounded soldiers. Another H-5 evacuated two more soldiers on the same mission. By mid-1951, USAF H-5s had airlifted 1,722 wounded in Korea, and the 3rd ARS became the first Air Force unit awarded a Distinguished Unit Citation.

The dangerous business of Combat Search And Rescue (CSAR) was still in its infancy. In March 1951, now-Capt. Miller flew his H-5 through intense fire to rescue a Marine Corsair pilot downed near Munsan and was awarded a Silver Star. Sikorsky S-51s would remain in Korea throughout the war and in Air Force squadrons until 1962.

S-55s and (briefly) S-58s

Air Force orders totaled 131 S-51s, but Air Rescue Service capability grew significantly with the Sikorsky S-55 (SH-19A and -19B). Two YH-19 prototypes joined the 3rd ARS in Korea. The first arrived on March 23, 1951 and carried wounded soldiers the next day. The July 17, 1951 *Sikorsky News* quoted visiting Air Force Capt. Richard McVay saying, "We want more H-19s over there." McVay reported, "Ninety per-cent of our behind-the-lines pickups of crashed pilots have been successful. We have made 463 such rescues behind the enemy lines."

The S-55/H-19 in Korea rescued downed pilots and airlifted wounded soldiers throughout the war.

The S-51/H-5H with floats, hoist, and cabin litter provisions remained in Air Force service until 1962.

The Air Force Reserve operated S-58s (HH-34Js) awaiting turbine-engine helicopters (Hugh Alcock)

One notable combat rescue came on April 12, 1953 when fighter pilot Captain Joseph McConnell claimed his eighth MiG over Korea but felt his Sabrejet falter over enemy territory. McConnell radioed for help and ejected over the Yellow Sea. Two H-19s from the 3rd ARS launched from Chodo Island and recovered the ace minutes after he hit the water.

The S-55 SH-19 rescue helicopters deployed with Air Rescue Squadrons around the world.

The Air Force ultimately acquired 320 S-55s deployed around the world. In July, 1952, two H-19s flown by Capt. Vincent McGovern and Lt. Harold Moore crossed the Atlantic from Westover Air Force Base, Massachusetts to Prestwick Scotland enroute to Wiesbaden, Germany. *Sikorsky News* reported, "Two days after arrival in Wiesbaden, the two H-19s rescued the crew of an American bomber that had crashed in the Rhine." In April 1954, H-19s of the 59th ARS hauled 30,000 lb of food to flood refugees in the Tigris River Valley of Iraq. That same month, an H-19 of the 56th ARS hoisted 18 men from a Swedish freighter run aground near Casablanca, Morocco. On 19 October, 1959, Captain Herbert Mattox flew an H-19 of the 33rd ARS to hoist 29 crewmen from the grounded Japanese vessel Zenko Maru and earned the 1959 Cheney Award.

Sikorsky Stratford opened in April 1955 and made S-58s (military H-34s) for the U.S. Navy, Marine Corps, and Army, and for global customers until 1970. The Air Rescue Service became the Aerospace Rescue and Recovery Service in 1966 and flew 32 retired Navy Seabats as HH-34Js from 1971 to 1974 in Air Force Reserve units awaiting turbine-engined helicopters.

Jolly Green Giants

Soon after the first twin-turbine S-61A (HSS-2 or SH-3A) Sea King flew on March 11, 1959 for the U.S. Navy, Sikorsky proposed a stretched S-61R with a rear ramp unsuccessfully to the Marines. However, the Air Force needed helicopters to shuttle to and from Atlantic radar towers, recover target drones, and service missile sites. The S-61R/CH-3C flew in June 1963, and went to Vietnam with the Tactical Air Command (TAC) in 1965. *Sikorsky News* in January 1965 carried a story on rescue HH-3Cs in Florida standing ready during Gemini space launches. The June issue reported, "Authorization also was received by Sikorsky for a special rescue mission configuration (with armor, self-sealing fuel tanks, and rescue hoist) for delivery to the Air Rescue Service."

The S-61R/HH-3E Jolly Green Giant with air refueling capability extended rescue range in Southeast Asia.

The Air Rescue Service borrowed two CH-3Cs from TAC in July 1965 and assigned them to the 38th ARS at Udorn, Thailand. The helicopters were quickly camouflaged in green paint and dubbed Jolly Green Giants, popular imagery from canned vegetable ads. In November 1965, the 38th ARS received two HH-3Es with 1,500 shp T58-GE-5 engines, 1,000 lb of guns and armor, self-sealing internal fuel tanks, and sponson drop tanks.

Armed Jolly Greens flew with four crew -- pilot, co-pilot, flight engineer, and pararescueman, or PJ. HH-3E pilot Lt Forrest Kimsey earned the Silver Star for efforts to recover a Navy Phantom crew down in a very 'hot' location near Tchepone, Laos on April 22, 1966. Jolly Greens operated in flights of two – a Low Bird to make the rescue and a High Bird orbiting in case the Low Bird was downed. The Low Bird leading Kimsey's flight picked up one aviator, but when the helicopter tried to recover the second, groundfire knocked the PJ off the hoist jungle penetrator and badly wounded the flight engineer. Kimsey's High Bird withdrew under heavy fire but refueled at the Khe San firebase and returned to the scene, again under fire, to pick up the second fighter crewman. Kimsey yet again returned at first light to recover the PJ.

By 1966, HH-3E detachments of the 37th and 38th Aerospace Rescue and Recovery Squadrons (ARRS) covered Vietnam, Laos, and the Gulf of Tonkin. A CH-3C with trial air refueling probe made dry hookups with a Hercules tanker in December 1965. First fuel transfer between an HH-3E with extendable probe and the new Air Force HC-130P Hercules tanker took place in December 1966, and air refueling was used in combat in 1967.

HH-3Es rescued 92 aircrew in 1966, 122 in 1967, 163 in 1968, 72 in 1969, and 47 as the Jolly Greens were phased out of the theater in 1970. In a single mission on March 30, 1968, four HH-3Es rescued 14 Marines from six helicopters shot down in South Vietnam.

The Air Force ultimately flew 50 HH-3Es around the world. In October and November 1969, Jolly Greens from the 58th ARR Squadron at Wheelus Airbase, Libya saved 2,516 Tunisians from floodwaters. On April 16, 1979, an HH-3E from Osan Air Force Base in Korea rescued 24 people from a sinking vessel. The Air National Guard received HH-3Es in 1975 and continued to operate them until the arrival of HH-60Gs. The last Special Operations MH-3Es deployed in Operation Desert Storm in 1991 and were retired in 1995.

Air Force HH-3Es served in the Air National Guard until replaced by the HH-60G Pave Hawk.

MH-3Es of the 71st Special Operations Squadron were CSAR and Special Warfare assets in Desert Storm. (Frank Colucci)

Super Jollies

Even with aerial refueling, the Jolly Green Giant was slow and struggled with high mountain rescues. *Sikorsky News* in March 1967 reported first flight of the new HH-53B rescue helicopter based on the S-65/CH-53A built for the Marine Corps. "The big, camouflage-painted helicopter carried auxiliary fuel tanks and a retractable probe for mid-air refueling as visible evidence of its long-range capability." The Super Jolly Green Giant arrived in Vietnam in September 1967. In January 1968, two Super Jollies recovered crews from an RB-66 jamming aircraft and a downed HH-3E. HH-53Bs of the 40th ARR Squadron made 99 combat saves in 1968. Super Jollies were soon the primary rescue helicopters in Southeast Asia. The interim HH-53B, with distinctive struts to support auxiliary sponson tanks, was soon replaced by the HH-53C, the final configuration delivered to the Air Force. In August 1970, two new HH-53Cs were ferried from Eglin Air Force Base, Florida to Danang, South Vietnam with air refueling in the first trans-Pacific helicopter flight.

In March 1972, five HH-53s recovered all 15 cr men of an Air Force AC-130 gunship shot down in Laos. Even with coordinated CSAR task forces, rescue missions were not without heavy losses. Super Jolly crews flew dangerous rescue missions through the end of the conflict in Southeast Asia

and evacuated personnel from Saigon in 1975. The U.S. Air Force bought 72 HH-53B/C helicopters. Select Super Jollies returning from the war were modified with longer-range fuel tanks and special equipment to recover space hardware with the 6594th Test Group at Hickam Air Force Base in late 1974. The aircraft had a secondary Search And Rescue mission, and on January 7, 1985, an HH-53 from Hawaii hoisted a seriously ill crew member from a distant cruise ship. The 1,380-mile round-trip marked a world record for the longest overwater helicopter rescue without a landing.

Super Jolly rescue helicopters made early use of primitive night vision equipment in 1972 and started the evolution of Pave Low night/adverse weather navigation capability. The first HH-53 rescue helicopters with low-light television and hover couplers were deployed to Nakhon Phanom, Thailand with limited success. The Pave Low program continued the evolution of night and adverse weather capability with the addition of Forward Looking Infrared sensors, terrain-following radar, and precision navigation capability. Initially, only nine HH-53H helicopters were planned for Pave Low conversion, but the success of the program led to all USAF HH-53s being modified. Pave Low HH-53H helicopters with improvements became MH-53J and -53M Special Operations aircraft flown around the world until their retirement in 2008.

The Sikorsky S-65/HH-53B Super Jolly Green Giant replaced the HH-3E in Southeast Asia.

The HH-53C went to war in 1969 and gave the Air Force a more powerful rescue/Special Operations aircraft.

Pave Hawk to Whiskeys

The Secretary of the Air Force approved a Mission Need Statement in November 1980 for a version of the Sikorsky S-70/UH-60A. *Sikorsky News* in December 1982 announced the Air Force contract to convert two Black Hawks into HH-60D Night Hawks. The same agreement also gave the Air Force 11 Army UH-60As from the Stratford line for pilot and maintenance familiarization. Sikorsky President Robert Daniel said, "The HH-60D will greatly enhance Air Force capabilities to conduct aircrew rescue operations deep behind enemy lines, in darkness or bad weather, and at treetop level, avoiding radar detection."

The Night Hawk flew for the first time on February 4, 1984 at the Sikorsky Development Flight Center in West Palm Beach, Florida and was actually credited with a save that September when it retrieved two hikers at 10,000 ft in the High Sierras near Bishop, California.

The Air Force upgraded the 11 Black Hawks to MH-60G Special Operations Pave Hawks with refueling probes in 1984, folding stabilators in 1985 and some special mission avionics in 1986. MH-60Gs flew in Operation Just Cause in Panama in 1989, and provided CSAR coverage for coalition forces in Operation Desert Storm in 1991. Further-modified HH-60Gs were based on the more powerful UH-60L and ultimately gave the Air Force 112 Pave Hawks for Rescue Squadrons (RQSs) starting in 1990.

The HH-60G Pave Hawk has been flown extensively in combat and relief operations. (U.S. Air Force)

The Pave Hawk evolved from the BLACK HAWK with air refueling and rescue avionics.
(U.S. Air Force)

Two HH-60G crews from the 56th RQS earned the 1994 Cheney Award for rescuing six sailors in extremely bad weather from a sinking tug off the coast of Iceland. During Operation Allied Force over Serbia in 1999, Pave Hawks helped recover two Air Force pilots down behind enemy lines. Operations Enduring Freedom in 2001 and Iraqi Freedom in 2003 put extraordinary demands on the helicopters. The 26th Expeditionary Rescue Squadron (ERQS) alone left Afghanistan in 2014 with 2,400 lives saved and more than 3,300 life-saving assists over five years, never with more than four Pedro HH-60Gs on-hand. Peacetime utilization was also high. After Hurricane Katrina in September 2005, more than 20 Pave Hawks flew in and around New Orleans to save more than 4,300 Americans. In August 2017, Pave Hawks from the 920th Rescue Wing reportedly saved 146 people in one day when Hurricane Harvey flooded Texas towns.

Attrition has trimmed the Air Force Pave Hawk fleet to 96 helicopters. The first of 21 Operational Loss Replacement aircraft converted from UH-60Ls to HH-60Gs for Air National Guard squadrons was accepted in June 2017. With greater range, truly integrated avionics, and enhanced digital connectivity, 112 new Sikorsky

HH-60W Combat Rescue Helicopters are to achieve Initial Operational Capability in 2020. They promise Air Force rescue crews new tools for their noble mission — *That Others May Live.*

Air Force Pave Hawks routinely operate from Navy Ships. (U.S. Air Force)

Here in final assembly, the Sikorsky HH-60W gives rescue crews greater range and new connectivity for Combat Rescue and Isolated Personnel Recovery. (Sikorsky, a Lockheed Martin Company)

The Aerospace Rescue and Recovery Service was disestablished in 1993, but the Pave Hawk remains in today's Air Combat Command, Air Force Reserve, and Air Guard. (U.S. Air Force.)

Sikorsky's Navy Subhunters

"In guarding pilot personnel of aircraft carriers by effecting a speedy rescue of pilots whose planes have plunged into the sea, the helicopter has virtually eliminated the need of use of destroyers with airplane carriers for a pilot safety surveillance."

Today's multi-mission, multi-sensor MH-60R gives the U.S. Navy the world's most advanced Anti-Submarine Warfare (ASW) helicopter. However, the potential of the helicopter in ASW was recognized even before the Navy and Coast Guard flew their first Sikorsky R-4s in World War II. In February 1943, the Chief of Naval Operations directed the Bureau of Aeronautics to develop the helicopter for shipboard anti-submarine patrol. Pioneer Coast Guard pilot Lt. Frank Erickson soon after proposed helicopters with radar and dunking or dipping sonar as the eyes and ears of maritime convoy escorts. Post-war experiments with a Sikorsky XR-6A (the Navy XHOS) showed helicopter sonar was indeed effective locating undersea targets. The very first newspaper edition of The *Sikorsky News* on October 12, 1948 pictured Rear Admiral James Fife, Commander of the Atlantic Fleet Submarine Force, visiting the Bridgeport plant ". . . to look into possibilities of [the] helicopter's utility in the spotting of submarines."

In April 1950, the Navy ordered utility S-55s (HO4S-1s with folding three-bladed main rotors and 550 hp Pratt & Whitney engines). When the tandem-rotor Bell Model 61 struggled with technical problems, the Navy chose the Sikorsky S-55 as its substitute sub hunter. Navy HO4S-3s had 700 hp Wright piston engines and could operate in ASW teams, the submarine "hunter" with dipping sonar and "killer" with a torpedo.

The S-55 (HO4S-3) gave the U.S. Navy its first operational ASW helicopter deployed on aircraft carriers.

The U.S. Navy commissioned its first antisubmarine warfare helicopter squadron, HS-1, at Naval Air Station Key West, Florida on October 3, 1951, and other navies followed. The UK Royal Navy formed 706 Squadron for ASW with Sikorsky-built S-55s in 1953, and Westland Helicopters began license production of the Whirlwind in the UK soon after. Canada formed Helicopter Anti-Submarine Squadron Fifty with S-55s in 1955.

The S-55 (HO4S-3) could carry either dipping sonar or a torpedo and had to work in hunter-killer teams.

Seabats

The 7,900 lb S-55 was short of range and payload for ASW. The April 21, 1954 *Sikorsky News* showed tie-down tests of the new S-58 (Navy XHSS·I), and the photo caption stated simply, "Specifications are classified." However, with a 1,525 hp engine and four-bladed rotor, the 14,000 lb S-58 (Navy HSS- 1 or SH-34G) Seabat promised to be a more effective hunter or killer, one even compatible with a nuclear depth charge. In September 1955, *Sikorsky News* reported on the successful HSS-1 test effort. Navy Squadron HS-3 was the first to receive the HSS-1 in 1955.

The S-58 (HSS-1) was designed for ASW. Here, a Seabat from HS-4 hovers with dipping sonar near the destroyer USS Richard B. Anderson in 1958. (U.S. Navy)

In 1956, Squadron HS-9 was commissioned at Quonset Point, Rhode Island. *Sikorsky News* reported "The new squadron is to operate from aircraft carriers of the Leyte class based at Quonset, working with other fleet units such as destroyers, submarines, patrol bombers and blimps."

However, without visual hover references at night, early Seabat crews could use dipping sonar only in daylight. The *Sikorsky News* in May 1958 told of developments in night hover capability. The new HSS-IN (SH-34J) combined Automatic Stabilization Equipment with an improved flight instrument and cockpit arrangement, automatic engine RPM controls, and an automatic hover-coupler. The story explained, "With the coupler, which uses the radar to determine ground motion, it is possible for the pilot to place the helicopter on automatic control at 200 feet altitude and 80 knots airspeed and automatically to come to a zero ground speed hover at a 50-foot altitude over a pre-selected spot."

Navy Squadron HS-5 at Quonset Point, Rhode Island received HSS-1Ns in 1960 and, after dangerous testing, the Nightdippers made autohover routine. The *Sikorsky News* reported delivery of the 1,000th S-58 in November 1958, an ASW helicopter made in Bridgeport for the U.S. Navy. More S-58 sub-hunters were produced under license by Westland in the UK, Fiat in Italy, and Sud Aviation in France. The March 1958 *Sikorsky News* reported, "Two S-58 helicopters have been purchased by the Japanese Navy for anti-submarine warfare use. . . [The S-58] has a payload capacity more than double the S-55 currently in use by the Japanese."

The HSS-1F testbed flew with twin GE T58 turboshafts in 1957.

Sea Kings

Despite its size, the Seabat was still limited by its single, heavy reciprocating engine. An engine failure would drop a sonar-dipping helicopter in the ocean, and though the S-58 could carry sonar and a torpedo at the same time, the weapon came at the expense of fuel and dramatically reduced endurance, especially in hot climates.

Sikorsky began investigating turbine power in the 1940s, and the S-59 (Army XH-39) set speed and altitude records in 1954 with a Turbomeca turboshaft. General Electric meanwhile proposed a small gas turbine specifically for helicopters. Twin GE T58s first flew in February 1957 in the first of two HSS-1F test aircraft. The March 1957 *Sikorsky News* reported "The new powerplant, developed for the Navy Bureau of Aeronautics, will replace the Wright R-1820."

When the Navy formulated a requirement for an ASW helicopter to carry sonar and weapons, it saw turbine power as an Engineering Change Proposal for the HSS-1. Sikorsky made its own proposal on January 31, 1957 for a new five-bladed, twin-turbine helicopter with a boat hull for overwater safety. The *Sikorsky News* revealed the 18,000 lb sub-hunter taking shape in Stratford and said, "The design

The S-61 mockup constructed at Bridgeport showed the distinctive Sea King boat hull.

also promises greater range, larger payload and better visibility than its predecessor, the HSS-1. First flight of the advanced helicopter is scheduled for 1959. Company designation of the new ship is S-61." A Navy contract was awarded on December 24, 1957, and what should have been the HS2S in service became the HSS-2 (later SH-3A) Sea King.

The second S-61 variant in the U.S. Navy (SH-3D) was the pattern for export Sea Kings.

The first of seven YHSS-2 prototypes flew on March 11, 1959 with company test pilots Robert Decker and Francis 'Yip' Yirrell. The new sub hunter soon broke records. On 17 May, 1961 a Sea King clocked 192.7 mph over a 3 km course, and on 1 December, another set three world helicopter speed records: 182.8 mph over a 100 km course, 179.5 mph over 500 km, and 175.3 mph over 1,000 km. An SH-3A reached 210.65 mph on a 20 km course in February 1962, and another, The Dawdling Dromedary, set a non-stop helicopter distance record flying 2,105.49 miles from the USS Hornet off San Diego to the USS Franklin D. Roosevelt off Jacksonville on March 6, 1965.

First Sea King deliveries began to squadron HS-1 in June 1961. The January 1963 *Sikorsky News* reported on Seventh Fleet operations: "U.S. Pacific Fleet Navy men are definitely excited about the new HSS-2 (SH-3A) helicopter. They believe this revolutionary turbocopter has opened a new chapter in the Navy's book on antisubmarine warfare." The story continued, "As the first Navy helo to combine both hunter and killer capabilities, the SH-3A can detect, identify, track and destroy aggressor submarines. It can accomplish this day or night, under all weather conditions."

Canada became the first international Sea King customer. The first four CHSS-2s (later CH-124s) left Stratford in May 1963. Subsequent helicopters were assembled from Sikorsky kits by United Aircraft of Canada Ltd (now Pratt & Whitney Canada) at Longueuil, Quebec.

The U.S. Navy SH-3A (S-61) evolved into the SH-3D in 1965 with more powerful engines, an uprated main gearbox, and a redesigned tail rotor. The SH-3D was first exported to Spain in June 1966 and provided the pattern for Sea Kings license-built by Westland in the UK, Agusta in Italy, and Mitsubishi in Japan. Westland ASW Sea Kings joined the UK Royal Navy in 1970 and were later sold to Australia and Germany.

The SH-3D introduced better dipping sonar, but the U.S. Navy Sea Control Ship concept in the early 1970s drove development of the multi-sensor SH-3H with sonar, Magnetic Anomaly Detector, sonobuoys, and search radar. The U.S. Navy converted 151 Sea Kings to SH-3Hs from 1972 to 1982, then recycled SH-3Hs through a Service Life Extension Program (SLEP) at Stratford beginning in 1986. Sea Kings remained in U.S. Navy service until 2009.

The SH-3H hunted with sonar, MAD, radar, and Electronic Support Measures.

Canada was the first international Sea King customer and went on to operate its CH-124s from frigate decks for 50 years.

Seahawks

The U.S. Navy put Kaman Seasprite helicopters on first-generation Light Airborne Multi-Purpose System (LAMPS I) ships in 1970. The helicopters datalinked sonobuoy returns to their ships for processing and used MAD to target submarines and radar to track surface targets.

The Navy wanted better sensors and more processing power on LAMPS II. *Sikorsky News* in April 1972 saluted the British WG-13 Lynx helicopter and reported, "Sikorsky is proposing the WG-13 as the Navy's LAMPS (light airborne multi-purpose system) vehicle and would produce the aircraft in this country under license from Westland."

LAMPS II was bypassed, and IBM Systems Integration Division (today Sikorsky Owego) won a Navy contract in 1978 to integrate LAMPS III systems on ships and helicopters. In the competition for the air vehicle, a marinized Sikorsky S-70/UH-60 Black Hawk again beat the Boeing YUH-61 designed for the Army Utility Tactical Transport Aircraft System. The new Seahawk was built in the then-new Development Manufacturing Center in Stratford. *Sikorsky News* reported pilots John Dixson and Richard Mills began testing the first of five S-70B-1 (Navy SH-60B) prototypes on December 12, 1979 at the Development Flight Center near West Palm Beach, Florida.

Sikorsky received a Seahawk production contract in September 1981. As LAMPS III co-prime contractor, the helicopter maker integrated SH-60Bs built in Stratford with mission electronics from Owego, New York. The first Seahawk squadron, HSL-41, was commissioned at Naval Air Station North Island, California in 1983. Worldwide, the 181 LAMPS III helicopters integrated at Stratford and Owego became powerful battle management assets for the U.S Navy.

The S-70B-1 or SH-60B was the airborne portion of the Light Airborne Multipurpose System Mk. III (LAMPS III) and fed sensor data to the ship for processing.

The S-70B-2 Seahawk was the Roll Adaptive Weapons System developed for the Royal Australian Navy.

The S-70C(M) Thunderhawk delivered to the Republic of China in 1991 evolved from the U.S. Navy S-70B- 4/SH-60F CV Helo.

In 1985 Sikorsky reported delivery of the first Seahawk for Japan. Mitsubishi Corporation integrated the S-70B-3 (SH-60J) with Japanese systems and later developed the SH-60J Kai in service today. A government-to-government sale delivered LAMPS III frigates and S-70B-1s to Spain in 1988. Unlike the highly integrated LAMPS III, the Royal Australian Navy specified a more autonomous Role Adaptive Weapons System. Sikorsky was prime contractor for the RAN S-70B-2 with Collins tactical avionics system delivered in 1989.

While the Outer Zone SH-60B used sonobuoys to find submarines, the noisy Inner Zone around the CV-class aircraft carrier demanded active dipping sonar for quick, autonomous attack. The Navy chose the S-70B-4 (Navy SH-60F) with dipping sonar, sonobuoy launcher, and more autonomous tactical data system. The "CV Helo" first flew at Stratford in March 1987 and entered service with HS-10 at North Island in June 1989. HS-2 on the Atlantic Coast made the first operational deployment in 1990. The Navy took delivery of the last of 82 SH-60F Seahawks on 1 December 1994. However, Sikorsky continues to develop international S-70Bs and has so far exported Naval Hawks to Brazil, Greece, the Republic of China, Thailand, and Turkey.

Naval Hawks and Cyclones

The SH-60B and F were designed for "Blue Water" ASW. However, Navy strike groups today face quiet submarines in the open ocean and cluttered, shallow littoral regions. The LAMPS III Block II Upgrade evolved into the more autonomous MH-60R with Advanced Low Frequency Sonar, Automatic Radar Periscope Detection and Discrimination, multi-spectral electro-optics, and better Electronic Support Measures (ESM). Engineering and Manufacturing Development started in 1993, and

The MH-60R Sea Hawk is the U.S. Navy multi-mission helicopter with sonar, sonobuoys, search radar, electro-optics, and ESM. (U.S. Navy)

Romeo Seahawks flying from carriers and small decks ultimately replaced the SH-60B, SH-60F, and fixed-wing ASW aircraft.

The first MH-60R – a remanufactured SH-60B -- made its maiden flight July 19, 2001 at Stratford with Sikorsky pilots Chris Geanacopoulos and Tage Erickson. After flight tests at Naval Air Station Patuxent River, Maryland, it flew to Owego for mission systems integration. The first operational MH-60R deployment was with Helicopter Maritime Strike squadron HSM-71 on the USS Stennis in 2009. *Sikorsky Frontlines* in 2010 reported the delivery of the 50th MH-60R. By 2018, the U.S. Navy had received 251 MH-60Rs. The Royal Australian Navy was the first international Romeo operator and took 24 aircraft. Romeo deliveries continue to Denmark and Saudi Arabia.

Australia was the first international customer for the MH-60R. (Australian DoD)

Separate from the Naval Hawk family tree, Sikorsky built 28 S-92 (CH-148) Cyclones for the Canadian Forces with a mission suite integrated in Canada. Test Pilots John Armbrust and Rick Becker made the first Cyclone test flight on November 14, 2008 at West Palm Beach. The militarized, marinized, fly-by-wire S-92s are assigned to the 423 Maritime Helicopter Squadron at Shearwater, Nova Scotia. Canada retired the Sikorsky Sea King in 2018 after more than a half-century of subhunting service.

The first CH-148 Cyclone Canadian Maritime Helicopter-flew on November 14, 2008 at West Palm Beach. (Sikorsky)

Canada now operates the CH-148 Cyclone, the marinized, militarized, fly-by-wire Sikorsky S-92. (Canada DND)

Sikorsky Builds Marine Corps Heavy Lift

"I forsee the creation of vast new detachments of air cavalry capable of landing large groups of men behind the fighting lines."

Sikorsky Aircraft made the U.S. Marine Corps vision of Vertical Envelopment real with progressively more capable helicopters.

Five years after the U.S. Marine Corps first practiced helicopter assault from the sea with S-51s (HO3S-1s), Sikorsky Aircraft flew a heavy lifter that started helicopter production at Stratford and the evolution of Marine expeditionary lift. The S-56 or XHR2S-1 flew for the first time at Bridgeport, Connecticut on December 18, 1953 with company pilots Jimmy Viner and Jim Chudars. It was the first twin-engined helicopter built for the U.S. military and the first with retractable landing gear, power-folding rotors for shipboard operations, and autostabilization equipment. The S-56 promised to concentrate, move, and sustain combat power from the sea. Though the big, fast Deuce never met Marine expectations, it led to heavy lift Sea Stallions, Super Stations, and today's King Stallion.

The S-56 or XHR2S helicopter flew for the first time at Sikorsky's Bridgeport, Connecticut facility in December 1953.

In September 1946, the commander of Fleet Marine Forces Pacific, General Roy Geiger, witnessed the first atomic bomb detonation at Bikini Atoll and reported the threat such weapons posed to large assault fleets. A special board of Marine officers considered alternatives. In February 1947, a report on Military Requirements of Helicopters for Ship-To-Shore Movement of Troops and Cargo described an assault helicopter to carry 15 to 20 Marines and a bigger aircraft to haul combat equipment. Col. Edward Dyer visited Bridgeport and outlined the need for a helicopter to carry 5,000 lb. Enthusiastic Igor Sikorsky told him, "We can build an airplane [he-

Igor Sikorsky was on hand for the first XHR2S flight by Jimmy Viner and Jim Chudars.

licopter] that will carry much more than that. We know how to do it. Take my word for it."

The Sikorsky S-51 (HO3S-1) first delivered to the Marines in February 1948 carried a useful load around 1,450 lb. Marine Helicopter Squadron HMX-1 that year used five of the light observation helicopters to shuttle 66 Marines and a few hundred pounds of equipment from the aircraft carrier U.S.S. Palau to Camp Lejeune, North Carolina. In 1950, HO3S-1s flew combat medical evacuation and resupply missions with Marine Observation Squadron VMO-6 in Korea. The Sikorsky S-52 (HO5S-1) replaced S-51s in the light helicopter squadrons in 1952.

The S-51 (HO3S-1) was used extensively in the Korean conflict

Bridgeport made S-55s for all the U.S. armed services Here, the line is full of HRS-2s for Marine medium lift squadrons.

With their objective heavy lifter still years in the future, the Marines ordered the Sikorsky S-55 (HRS-1) in 1950 and commissioned their first Medium Helicopter Squadron in 1951. HMR-161 took HRS-1s to Korea and began operations in support of the 1st Marine Regiment in August 1951. In successive operations, HRS-1s and -2s moved troops and rocket launchers, and by the armistice, HMR-161 had flown more than 18,000 sorties demonstrating Marine air mobility in combat.

The medium lift S-55 or HRS-1 gave the Marines air mobility in Korea with HMR-161.

The Sikorsky S-52 or HO5S-1 flew Medevac and resupply missions with Marine Corps squadron VMO -6 in Korea.

Marine Helicopter Squadron HMX-1 was the test unit for helicopter development. Here, an S-55 or Marine HRS flies with floats for water landings.

The 1,000th S-55, a Marine HRS, is inspected by Igor Sikorsky and company general manager Bernard Whelan.

The S-58 or HUS-1 was the Marine medium lift helicopter early in the Vietnam war.

In May 1951, Sikorsky won a contract for the S-56 (HR2S-1), a powerful, long ranged, instrument flight capable helicopter to insert Marines behind the enemy and exploit ter- rain and surprise with "vertical envelopment." The challenges of a marinized heavy lift helicopter able to fly at night and in adverse weather led the Marines to order the 14,000 lb Sikorsky S-58 (HUS-1/CH-34) in 1954. The Marine HUS-1 entered service in February 1957. Operation Shufly first sent Marine Medium Lift Squadron HMM-362 to South Vietnam with CH-34s in 1962. The Marines ultimately bought more than 600 S-58s, and the hard working HUS was not retired altogether until 1972.

Deuces Wild

As an assault helicopter, the 31,000 lb XHR2S was big enough for 26 Marines -- two combat assault squads -- or 24 casualty litters. Nose doors let jeeps drive in and out. A monorail crane carried a ton of cargo from clamshell nose to aft side door, and a belly hook could carry 10,000 lb sling loads. Sikorsky initially proposed a traditional S-56A with piston engines and a compound S-56B with turboshaft propulsion, thruster or tractor propellers, and lifting wings. With jet engines in short supply, the Marines bought the conventional helicopter with five-bladed main rotor and 1,900 hp Pratt & Whitney R-2800 radial piston engines like those in Corsair fighters of World War II.

Sikorsky's new 250-acre home in Stratford, Connecticut was dedicated on March 26, 1955.

Marines board an S-58/UH-34D aboard the aircraft carrier USS Bennington in 1963. (US Navy)

Sikorsky mocked up the S-56B with pusher turboprops and folding wings.

production brought previously outsourced manufacturing under the new roof. Five-pointed magnesium rotor head stars forged in Massachusetts, for example, became the first hub plates precision-machined, inspected, polished, and corrosion-coated in-house.

The HR2S-1 lifts a Sikorsky HUS (S-58) in a downed aircraft recovery exercise.

Much of the S-56 workforce was in place by May 11,1956 when Marine Major Roy Anderson and Sikorsky assistant chief of flight test Robert Decker flew from Stratford to Naval Air Test Center Patuxent River, Maryland to deliver the first HR2S-1. The 236 mile flight lasted 2 hours, 36 minutes and averaged 101 mph ground speed. On November 10, 1956, Anderson and Sikorsky pilot Robert Duke flew an HR2S-1 to 7,000 ft with a 13,250 lb payload and 12,000 ft with an 11,050 lb payload, breaking a record set by the Soviet tandem rotor Yak-24. Nine days later, an empty HR2S-1 set a new world's helicopter speed record at 162.7 mph (142 kt).

By the time the first HR2S-1 was delivered for testing by HMX-1 at Quantico, Virginia, the big helicopter was in production configuration with 2,100 hp engines, a 72 ft diameter main rotor, and other refinements. However, S-56 automatic stabilization equipment with attitude and heading hold was unreliable. Intense vibration took a toll on cockpit instruments. Reciprocating engines guzzled oil and needed frequent overhaul. Hydraulic rotor folding was troublesome.

The HR2S flew its shipboard trials aboard the USS Tarawa in 1956 and deployed aboard ships to support the Carbbean Ready Force starting in 1962.

HMR(M)-462 was commissioned at Mojave California as the second and last Deuce squadron. The S-56 joined the Army as the CH-37A in 1956. Sikorsky delivered the 100th S-56, a Marine HR2S-1, in 1958 and the last of 55 production Deuces to the Marines in 1959. The Marine HR2S-1 became the CH-37C, and Deuce squadrons became Marine Heavy Helicopter Squadrons in 1962.

CH-37Cs went to sea with the Caribbean Ready Force and to war in Vietnam. Their primary mission in the combat theatre was hauling troops, food, and ammunition internally. By May 1967 when "Junkman" operations ended in Vietnam, Deuces had moved 32,000 passengers and 12.5 million pounds of cargo.

Sea Stallions

The first operational CH-53As were delivered to HMH-463 on September 12, 1966 for the Fleet Introduction Team.

When the Navy withdrew from tri-service XC 142 tilt-wing turboprop development, the Marines sought a CH-37C replacement with power to move heavy cargo and recover downed aircraft. Turboshafts with high power-to-weight ratios promised to revolutionize helicopter performance. The Navy sponsored development of the General Electric T64 turboshaft/turboprop first

run in 1959. Sikorsky won the Heavy Helicopter Experimental competition in August 1962 with the S-65 (CH-53A).

A CH-53A sling-lifts a 105 mm artillery piece to An Hoa, Vietnam during Operation Sussex Bay, summer 1968. (US Marine Corps)

Marine CH-53As went to Vietnam in 1967. Here, one delivers barbed wire to protect the Marine outpost at Con Thien.

The CH-53D with -413 engines upped the Sea Stallion gross weight to 42,000 lb.

CH-53D production delivered 124 improved Sea Stallions to the Marines, plus export aircraft for Israel and Austria. CH-53Gs were built in Germany.

The new 35,000 lb helicopter had two T64-GE-3 engines to drive a six-bladed main rotor, 72 ft in diameter. Though largely aluminum structure like the S-56, the S-65 made extensive use of titanium in place of heavier steel for dynamic and structural components. *Sikorsky News* noted CH-53A hub plates were the largest titanium forgings used to date in helicopters and reported the introduction of numerically controlled milling machines and chemical milling at Stratford. The CH-53A also made early use of composite materials with fiberglass cockpit canopies in place of metal.

The first five production helicopters with 2,850 shp T64-GE-6 engines were delivered to HMH-

463 at Santa Ana, California in September 1964 for a Fleet Introduction Team. The squadron took CH-53As to Vietnam in January 1967, and by the end of the year, the 36 helicopters in theater had retrieved 370 downed aircraft. The Marines ultimately received 138 CH-53As.

The baseline S-65 was powerful, fast, and agile. On October 23, 1968, a CH-53A flown by Marine Lt. Col. Robert Guay and Sikorsky pilot Byron Graham flew loops and rolls over Long Island Sound. The first CH-53D with 3,925 shp T64-GE-413 engines, uprated transmission, seats for 55 troops, and other improvements was delivered to the Marines in March 1969. In May 1970, Sikorsky began work on S-65 Improved Rotor Blades (IRBs) with titanium spars bonded to aramid honeycomb and wrapped in fiberglass. The wide-chord IRB with new airfoil and twist hiked the CH-53D gross weight to 42,000 pounds. In 1972, Sikorsky introduced an elastomeric rotor head that eliminated oil lubrication and cut parts count by 30%. Elastomeric hubs were retrofitted to in-service CH-53Ds, T64 engines were uprated to 4,100 shp -416 and finally 4,500 shp -419 standards.

The CH-53D carried Marines and cargo in Afghanistan until it was retired in 2003. (US Navy)

The CH-53E used the power of three T64 engines and seven main rotor blades to haul heavy cargo for the Marines.

The CH-53E Super Stallion entered service in 1981 and remains in eight active and one reserve HMH squadrons.

The CH-53A and -53D served widely with Marine squadrons on land and assault ships. The Marines ultimately received 124 CH-53Ds. Delta model Sea Stallions hauled evacuees from Saigon to ships in 1975, carried air assaults in Desert Storm, and moved Marines in Afghanistan until the last retired with HMH-363 in 2012. The air-refuelable S-65 also spun off U.S. Air Force rescue and Special Operations helicopters and Navy minesweepers. S-65C export versions still serve Germany, Israel, and Iran.

Super Stallions

In late 1970, the Marine Corps broke with joint-service Heavy Lift Helicopter plans and received Congressional approval for a three-engined, seven-bladed stretch of the twin-engined, six-bladed CH-53A/D. The CH-53E (S-65/S-80) Super Stallion with three T64-GE-416 engines was to carry 16-ton external loads to support amphibious assaults and recover downed aircraft, including the CH-53D. It would more than double the external load of the CH-53D in a deck footprint just 10% larger than its predecessor. The Navy simultaneously wanted a heavy-lifter for Vertical On-Board Delivery to aircraft carriers. In January 1973, *Sikorsky News* quoted Naval Air Systems Command project manager Colonel

F. M. Kleppsattel, "The CH-53E will solve most of the heavy lift requirements of the Marine Corps and Navy during the foreseeable future."

The CH-53E production contract came on February 27, 1978, and the first production helicopter joined HMH-464 at New River, North Carolina in 1981. Composite medium lift squadrons are reinforced with heavy lift detachments when deployed. HMM-162 was the first to take CH-53Es overseas when it went to Beirut in 1983. The Super Stallion became essential to full-spectrum Marine operations. In September 1989, HMH-461 deployed CH-53Es to Puerto Rico for Hurricane Hugo relief. In 1990 and 1991, Operations Desert Shield and Desert Storm saw CH-53Es of HMH-464, -465, and -466 move Marines and cargo to liberate Kuwait.

CH-53Es extended the reach of MEU (SOCs) [Marine Expeditionary Units (Special Operations Capable)]. Marine Major, later Sikorsky President, Dan Schultz, led two CH-53Es on a 900 nm round-trip from USS Trenton to the besieged U.S. embassy in Mogadishu, Somalia during Operation Eastern Exit in 1990. The helicopters inserted Navy SEALS and Force Recon Marines to secure the compound and extracted 61 multi-nationals to Navy ships with three air refuelings – two at night. Five years later, CH-53E crews rescued a downed Air Force pilot from Bosnia and returned him to the USS Kearsarge.

The CH-53K Heavy Lift Replacement helicopter was de-signed to carry about three times the load of the CH-53E yet fit the same deck footprint. (US Navy)

In Operations Enduring Freedom and Iraqi Freedom, and subsequent actions, Super Stallions provided the power to move and resupply forces at high density altitudes. Sikorsky delivered the last of 172 CH-53Es in 2003. The Marine fleet has since received night visionics and navigation, communications, and survivability upgrades. The Marines now expect to "sundown" the CH-53E in 2032 when they achieve full operational capability with the new CH-53K.

King Stallions

The Sikorsky S-95 or Marine CH-53K Heavy Lift Replacement helicopter started as a low-risk derivative of the CH-53E. It emerged as an all-new technology helicopter to fit the Super Stallion deck footprint yet carry nearly three times the payload to high and hot landing zones. The 88,000 lb King Stallion was also designed to enhance survivability and cut operating costs compared to its predecessor, and to provide digital connectivity in a networked battlespace.

Marine requirements called for a Heavy Lift Replacement helicopter able to haul 27,000 lbs over 110 nm high and hot. The King Stallion has three fuel-efficient 7,500 shp General Electric T408-GE-400 turboshafts to drive a split-torque gearbox that packs 30% more power in a CH-53E-sized transmission. To fit assault ships, the CH-53K main rotor matches the 79 ft diameter rotor of the '53E, but wider-chord,

The CH-53K line in Stratford is filling with low rate initial production.

The CH-53K undergoing air to air refueling tests behind a Lockheed Martin KC-130.

drooped-tip, fourth-generation all-composite rotor blades generate 30% more lift and ride a titanium hub designed for on-condition maintenance.

King Stallion main rotor blades are the largest all-composite helicopter blades ever made by Sikorsky with fiberglass skins covering honeycomb cores and graphite spars 30 ft long. By wetted area, the CH-53K is more than 75% fiber reinforced composites. Fly-by-wire flight controls optimize CH-53K handling with heavy loads and in degraded visual environments. An integrated "glass" cockpit manages crew workload.

Sikorsky received the CH-53K System Development and Demonstration contract in April

2006 and opened its Heavy Lift Development Center in Stratford in February 2007. King Stallion parts were designed, fabricated, and integrated in a virtual environment that accelerated assembly of the real helicopter.

Stratford facilities static tested the CH-53K airframe, whirl-tested King Stallion rotor blades, and integrated the glass cockpit of the new heavy lifter. A new 60,000 sq ft hangar at the Sikorsky Development Flight Center (DFC) in West Palm Beach, Florida opened in 2012 to assemble and house King Stallion test aircraft. Four CH-53K Engineering Development Models (EDMs) and a fully functional Ground Test Article were assembled in Florida.

EDM-1 hovered for the first time at West Palm Beach on October 27, 2015. CH-53K Low Rate Initial Production aircraft are on the production line today in Stratford. The Marine Corps plans 200 King Stallions to equip eight active duty squadrons, two reserve units, and one training squadron, and Sikorsky's latest heavy lifter is competing for international orders.

The Marine Corps is refocusing from fighting extremism in the Middle East to peer-level conflicts with emphasis on the Indo-Pacific region. Warfighting scenarios are turning from inland to littoral battlespaces. The joint service Future Vertical Lift initiative may give future Marines a fast, long ranged Attack Utility Replacement Aircraft (AURA) based on Sikorsky X2 technology. Whatever and wherever the conflict, the expeditionary Marines will still depend on vertical lift. Sikorsky helicopters were the first to give them forceful, flexible lift from the sea, and they will remain central to the Marine Corps' evolving mission.

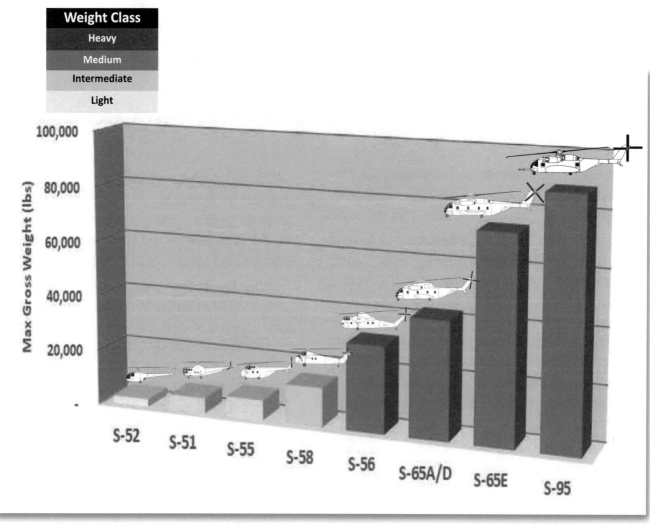

The maximum gross weight of Sikorsky helicopters for the Marine Corps has increased by over 85,000 pounds since the first delivery of the S-52.

Chapter 5
Commercial Operations

S-76

S-92

Sikorsky Helicopters in Airline Service

"Los Angeles Airways started an airmail service using S-51 helicopters on October 1, 1947. This was the first helicopter airmail service in the world. They also poineerd routes that led to ... passenger service in 1954."

Early in the evolution of Sikorsky helicopters, Igor Sikorsky and his team envisioned scheduled passenger flights serving major cities. The December 10, 1948 *Sikorsky News* told of a heliport survey flown by chief pilot Jimmy Viner in and around New York. "This futuristic and objective study was made in a S-51 helicopter with three-fold purpose: to plan for the inevitable air-mail helicopter pick-up; to establish passenger-to-airport shuttle service; and to plan for landing places for helicopters, which already travel to and from the city with passengers." The four-seat S-51 flew from Sikorsky's Bridgeport, Connecticut plant to the 36th Street pier on the Hudson River

A Sikorsky News story in 1944 reported Greyhound president Arthur Middleton Hill foresaw a helicopter-bus network serving 1,100 American towns. An S-51 was used to demonstrate the concept years later.

and on to New York International (today JFK International), Newark Liberty International, and LaGuardia Airports, routing over waterways to comply with local ordinances.

Sikorsky S-51s began airmail services in Los Angeles in 1947 followed soon after by Chicago and New York. The subsidized airmail runs defined paying passenger routes. When the YH-19/S-55 first flew in 1949, Sikorsky quoted a price of $120,000 for a passenger-carrying version. The commercial S-55A was certificated by the Civil Aeronautics Authority (now the Federal Aviation Administration) in 1952. Los Angeles Airways (LAA) ordered the first two helicopters in 1951 and began twice-daily mail runs between Los Angeles and San Bernadino in June, 1952. Passenger flights from Los Angeles International Airport (LAX) to Ontario and San Bernardino started December 17, 1953. By April 1957, LAA had 17 weekday departures from LAX to 11 area heliports.

The December 19, 1952 *Sikorsky News* reported on a study commissioned by the Port of New York Authority predicting, "Helicopters will carry over six million intercity, suburban and aerocab passengers, over 40 million pounds of airmail, and over six and one half million pounds of package cargo a year into and out of New York City by 1975." Mail-carrying New York Airways (NYA) began space-available passenger service with the eight-seat S-55 on July 8, 1953 and by the end of the year had routes stretching from Bridgeport, CT to Trenton, NJ. covering the metropolitan airports. That December, *Sikorsky News*

Los Angeles Airways began S-51 heli-mail service in 1947, defining routes to be used for scheduled passenger service.

announced New York Airways had a ticketing, baggage handling, and advertising agreement with United Airlines (UAL) and reported, "NYA President Robert L. Cummings said the arrangement with UAL would solve one of the copter line's greatest problems: getting itself known to the public."

SABENA World Airways began S-55 passenger service in 1953.

New York Airways bought four S-55s for $150,000 each in January, 1953. By 1957 the helicopters were flying from their Hudson pier heliport to LaGuardia Airport in 11 minutes for $5, Newark Airport in 12 minutes for $6, and Idlewild (now JFK) in 20 minutes for $7. National Airlines briefly served the Miami area with S-55s in 1953 and 1954, and Mohawk Airlines shuttled S-55s from New York to the Catskills in 1954. Chicago Helicopter Airways began S-55 passenger service in November, 1956 and soon introduced the 12-seat S-58.

In Europe, SABENA World Airways of Belgium took the lead with four S-55s, starting the first international helicopter passenger service in September, 1953. The service upgraded to eight S-58s in 1957. By 1960, SABENA helicopters connected Brussels with Antwerp, Rotterdam, Paris, Lille, Eindhoven, Maastricht, Liege, Dortmund, Duisburg, Bonn, and Cologne.

SABENA upgraded to S-58s.

Chicago Helicopter Airways S-58.

The November 20, 1953 *Sikorsky News* reported on a talk by Eastern Air Lines engineering vice president Charles Froesch who envisioned 30- to 50-seat helicopters. Sikorsky considered passenger versions of the big piston-engine S-56 (Marine HR2S-1) first flown in 1954. However, the twin-turbine power of the S-61 (Navy HSS-2) flown in 1959 promised new speed and safety for passenger operations. Sikorsky unveiled a civil S-61 mockup with stretched cabin in 1959.

New York Airways put the twin-turbine S-61L in service in 1969.

The 28-seat S-61L 'cop-terliner' first flew on December 6, 1960, and *Sikorsky News* that month announced, "The S-61L is quieter, smoother, carries twice as many passengers and flies 30 percent faster (136 miles an hour cruising speed) than any helicopter presently serving a scheduled helicopter airline. Los Angeles Airways has ordered five of

LA Airways S-61 over Disneyland.

the new Sikorsky turbocopters and Chicago Helicopter Airways is purchasing four." Unlike the Navy's Sea King, the non-amphibious airliner had fixed landing gear without sponsons, a forward cargo door and rear passenger airstair on the right side, and baggage bays in the boat hull.

Los Angeles Airways took delivery of its first S-61L on January 7, 1962 and flew the airliner between LAX and Anaheim on a Disneyland Connection. New York Airways began turbine service with the 25-seat Boeing Vertol 107 in 1962, but traded up to the 30-seat S-61L-4 in 1969 with better One Engine Inoperative performance for the Pan Am Building heliport 900 ft above midtown Manhattan. Pan American World Airways provided the financial backing for NYA and actually placed the order for five S-61Ls in 1967.

SFO also operated the single-engine Sikorsky S-62.

Sikorsky also developed an FAA-certificated amphibian airliner for over-water routes. The S-61N flew on August 7, 1962 with float sponsons and retractable landing gear and a sealed hull with baggage compartments accessible from the cabin. The first S-61N customer was Pakistan International Airways. East Pakistan (today Bangladesh) Helicopter Service started commercial flights on November 25, 1963. The helicopter from Dacca to Khulna cut a 21-hour trip over land to just 37 air minutes. The airline ultimately served 20 towns and cities.

In October 1964, the S-61L and N became the first transport helicopters FAA-approved for Instrument Flight Rule passenger operations. An NYA S-61N was sold to San Francisco and Oakland Helicopter Airlines in 1965. SFO eventually had four S-61Ns running between the San Francisco and Oakland airports and connecting San Francisco with Sausalito, Berkeley, and the Emeryville heliport. *Sikorsky News* in December 1970 reported "Sikorsky Aircraft's S-61N and S-61L commercial helicopters, flown on regularly scheduled routes, have passed the four million passenger mark."

Greenlandair, Ansett ANA in Australia, Elivie in Italy, and other international operators used S-61Ns in scheduled services. British Caledonian Helicopters ran an S-61 Airlink between Gatwick and Heathrow airports from 1977 to 1987. British European Airways Helicopters received two S-61N-2s in 1964 and began scheduled passenger service between Penzance and the Scilly Islands archipelago 28 miles off the southern tip of England. BEA Helicopters became British Airways Helicopters in 1974 and in 1986 was sold to a new owner who continued to fly two S-61Ns as British International Helicopters (BIH). BIH carried the four-millionth passenger on the Penzance run in 2010. The airline logged 90,000 passengers a year and 12 flights per day by 2011 but ended with a real estate dispute in 2012.

The vision of helicopter airlines drove Sikorsky advanced concepts. A study by the Los Angeles Department of Airports and the U.S. Department of Housing and Urban Development, for example, considered a Skylounge. *Sikorsky News* in December 1966 explained, "The unique system proposes to use a 40-passenger vehicle to pick up passengers and baggage at points in downtown Los Angeles. It then would be driven to a central pickup station where a crane-type helicopter would airlift the vehicle, passengers and baggage directly to Los Angeles International Airport. . . Sikorsky's role in the study will be to provide detailed information on performance of the S-64 crane-type, heavylift helicopter which will airlift the lounge."

S-61Ns flew scheduled runs to connect the south of England with the Sicily Islands for 48 years.

For all their promise, most helicopter airlines fell victim to business challenges. The December 1972 *Sikorsky News* reported on the long-forgotten Island Helicopter service from Garden City, Long Island to New York City heliports with single-turbine S-62s. U.S. Helicopter began twin-turbine S-76 service between Manhattan and New York airports in 2006 but shut down in 2009.

Significantly, Helijet in British Columbia continues the Vancouver-Victoria S-76 shuttle begun in November 1986. The helicopter airline expanded commuter ser-

Helijet has operated scheduled S-76 services between Vancouver and Victoria British Columbia since 1986 and recently expanded commuter service to Nanaimo

vice to Nanaimo in March 2015. By October 2017, Helijet had carried 2.26 million passengers. Vice president Rick Hill credited location and demographics for sustained success. "Vancouver BC is the largest city in British Columbia, and Victoria BC is the Provincial Capital of BC located on Vancouver Island, separating the two by a large body of water, the Georgia Straight. With a high propensity of business travel between the two cities and the convenience of being able to offer quick and efficient service to/from the city cores, the helicopter is a valuable tool for the busy Government and Business sectors." He added, "The S-76 aircraft offers the smooth, fast and reliable attributes that are expected by the business community. With its 12-seat capacity, it is the right size for the mission and offers the speed and comfort needed to deliver a high-frequency commuter-type operation." Helicopter airlines remain viable only on niche routes not easily served by fixed-wing aircraft. By design, Sikorsky's successful S-92 was sized and equipped for oil rig support, executive transport, search and rescue, military transport, and commuter airlines. Most of the fleet serves in high-intensity scheduled charter operations. However, when the S-92 prototype first flew on December 23, 1998, it carried a clear if now-forgotten name -- Helibus.

Sikorsky S-92 departs the Downtown Manhattan Heliport.

Sikorsky Strikes Oil

"Most interesting results have been achieved by some non-scheduled operators. The remarkable work of the oil field operations in the Gulf of Mexico must be regarded as a significant and promising beginning of a great and very important new type of transportation. In general, there is an interesting and most promising future for the helicopter."

Norsk Helikopter received its first Sikorsky S-92 in 2005. In 2006, it became the first S-92 North Sea operator to attain 10,000 fleet flight hours. (All images property of Igor I. Sikorsky Historical Archives)

Early in their evolution, helicopters became important tools in the global oil and natural gas industry. Petroleum Bell Helicopters first used light helicopters in 1949 to fly survey teams around southern Louisiana, and by the end of that year eleven oil and gas fields were identified in the Gulf of Mexico. In March 1952, the 7,200 lb Sikorsky S-55 became the first large commercial helicopter certified by the U.S. Civil Aeronautics Authority. The January 1954 *Sikorsky News* reported an S-55 demonstrator had toured Louisiana and Texas to acquaint oil company executives with the new helicopter for offshore operations.

demonstrator had toured Louisiana and Texas to acquaint oil company executives with the new helicopter for offshore operations.

Sikorsky Archives President Dan Libertino joined the renamed Petroleum Helicopters International (PHI) in 1954 as an S-55 mechanic at Grand Isle, Louisiana. He recalled, "The S-55 was the aircraft of choice for offshore oil operations because you could get five or six people on board." Humble Oil hired PHI to shuttle workers from Grand Isle to three offshore drilling rigs. "At Grand Isle, we had a hangar to get the aircraft out of the elements. We did all their inspections at night after the day's operation was over with." The oil company capped S-55 gross weight at 6,900 lb to operate from platforms built on war-surplus landing craft anchored beside drilling rigs and from the first helipad built on an oil rig.

A January 1955 article in the United Aircraft Corp. *Bee-Hive* magazine noted that, before helicopters, workers in the Gulf of Mexico spent up to six uncomfortable hours riding crew boats to and from their platforms. Two PHI S-55s flew 25 trips a day covering legs 7 to 42 miles long in just 6 to 40 minutes. Another helicopter stood by to keep operations on schedule. In addition to oil rig crews, the helicopters could deliver cargo or evacuate injured workers to shore. PHI acquired another S-55 from National Airlines, but a landing mishap in marginal weather halted the Humble Oil contract. "We were stuck with four or five S-55s," recalled Libertino. "Without having any work, we were at the point of taking cuts in pay or layoffs. All of sudden, PHI got a contract from Cities Service. Then we started to work for Gulf Oil." The S-55s reacquired pontoons for emergency water landings and set the stage for more capable helicopters.

The 14,000 lb S-58 was certified by the Federal Aviation Administration on August 2, 1956. Dan Libertino joined Sikorsky that November as a technical representative and soon travelled to New Guinea in the southwest Pacific with Worldwide Helicopters. Three S-58s helped develop drilling sites for Australasian Petroleum. "It wasn't practical to build roads," Libertino explained. "We were located at stations in the interior of New Guinea. All the material to make a clearing was broken down into 4,000 lb loads." With landing zones cut in the jungle by local workers, the helicopters hauled a bulldozer and supplies to successive worksites piece-by-piece. *Sikorsky News* reported two S-58s airlifted a million pounds in 45 days and noted that drilling equipment, which would have taken two years to deliver by truck if a road had to be built, was placed in three weeks by helicopter.

Westland in the UK licensed S-55 production in 1950, and in 1955 Bristow Helicopters began flying Westland Whirlwinds under contract to Shell Oil Company in the Persian Gulf. In 1957, Bristow expanded operations to Iran and Bolivia. Sikorsky ad copy meanwhile credited S-55s and S-58s in the Gulf of Mexico with more than 27,000 passengers a month. It continued, "Now in regular service in the Gulf of Mexico and other areas, the S-58 equipped with floats can fly

The piston-engined S-55 was the first transport category helicopter in offshore oil service, flying workers to rigs in the Gulf of Mexico in 1954.

The S-58 was equipped with 'hot-dog floats' for PHI to serve Humble Oil and Refining in the Gulf of Mexico.

The turbine-engined S-62 began offshore oil operations in the Gulf of Mexico in 1963.

125 miles offshore and return without refueling. It cruises at more than 90 miles per hour, making trips in only minutes which would take hours by surface craft. The S-58 is the industry's most advanced offshore transport." Sikorsky gave the piston-engined S-58 four floats and retractable wheels for amphibious operations, but a turbine-powered amphibian was on the way.

Turbine Times

The 8,300 lb S-62 with its General Electric CT58 turboshaft engine flew for the first time on May 14, 1958 and earned its FAA type certificate in June 1960 as the first aircraft to satisfy then-new regulations governing commercial passenger-carrying helicopters. The first production S-62 was delivered to PHI on July 23, 1960. *Sikorsky News* in November 1963 reported, "The turbine-powered Sikorsky S-62 has appeared in the Gulf and, with its boat hull (requiring no floats), has proved a natural for the offshore mission." Rotor Aids flew two S-62s and three S-55s from Grand Isle for Humble Oil. One rig 200 miles away could be reached in 105 minutes by S-62. *Sikorsky News* quoted Humble district superintendent James Walvoord, "The helicopter means added safety in cases of sickness and accident . . . We can operate from one base; without helicopters we'd need three or four bases."

The single-engine, 10-passenger S-62 captured few oil industry orders. Sikorsky first proposed airliner versions of the twin-turbine S-61 in 1959 and flew the 20,500 lb S-61N in August 1962 with a sealed hull and float sponsons for overwater operations. In May 1965, an Okanagan Helicopters S-61N made the first transatlantic crossing by a commercial helicopter flying 37 hours and 11 minutes in stages from Stratford to London to support Shell in then-new North Sea oil fields. British European Airways introduced the 26-seat

Okanagan began offshore operations with the twin-turbine S-61N for Shell Canada in 1968.

helicopter into the North Sea in 1965 followed by Dutch KLM and Norwegian Helikopter Service in 1966. Brunei Shell Petroleum started flying two S-61Ns in Borneo in 1967.

In the Gulf of Mexico, PHI tested an S-61N from April to August 1967, but only three platforms in the Gulf then had helipads big enough for the stretched S-61. Bristow flew S-61Ns in Malaysia in 1968 and started North Sea operations at Aberdeen, Scotland in 1971. A year later, the operator under contract to Shell Oil placed a single S-61N at Sumburgh to the south and ultimately launched 30 flights a day. By the mid-1970s Bristow was flying 18 S-61Ns, a turbine-powered S-58 Westland Wessex, and ten piston-engined S-58s from Aberdeen and Sumburgh averaging 100 hours per month per helicopter. The last S-61N built was delivered to Okanagan Helicopters in early 1980 and flew from Stratford to Nova Scotia, Canada for offshore work.

The S-64 Skycrane earned its FAA Type Certificate on July 30, 1965, and the big cargo lifter began a commercial sales drive including oil industry applications. *Sikorsky News* in January 1968 reported on a Skycrane demonstration conducted by Shell Oil. The 38,000 lb crane helicopter carried a complete sand-wash oil rig 11 miles from a marshalling yard at Leesville, Louisiana to an offshore structure in the Gulf of Mexico. In April 1969, ERA Helicopters purchased the first two commercial S-64Es for use in Alaska to transport heavy drilling equipment, but the powerful crane never became a regular feature of offshore operations. Erickson Air-Crane acquired the S-64 type certificate in 1992 and continues to offer heavy-lift services to the oil and gas industry.

Made for the Market

Sikorsky Light Twin Helicopter market studies in the early 1970s defined the company's first true

VOTEC Taxi Aéreo in Brazil began oil operations with the Sikorsky S-61N in 1979.

commercial helicopter, one tailored to the oil industry. The notional S-74 Twin Centurion became the S-76 Spirit, the name chosen from contest entries by Sikorsky president Gerald Tobias to tie in with the U.S. Bicentennial celebration. The Spirit of '76 was unveiled in February 1975 by Tobias at the Helicopter Association of America convention in Anaheim, California. Tobias championed the program before United Technologies management. "Without him it would never have happened," acknowledged Bill Paul, then S-76 chief engineer. Paul explained, "For the first time in our history, we developed a commercial helicopter from scratch. The others were using

Sikorsky performed an offshore heavy lift demonstration with an S-64 over the Gulf of Mexico in 1968.

a military base because it would be very difficult to amortize that [development] cost . . . All of the suppliers had to invest in it as well."

Informal talks with fleet operators had sketched a 10,000 lb helicopter with 400 nm range carrying 12 passengers and two crew at 145 kt cruising speed. Significantly, Category A performance requirements called for return to base at any point in the mission with one engine inoperative at maximum gross weight. "What Tobias added to that was the comfort of a corporate aircraft," noted Paul. "We recognized we wanted a market that would be broad enough, and Tobias was looking to the corporate market. In the end, it was the corporate market that saved our skin."

The elegant S-76 had a bifilar rotor head to suppress vibration and a main rotor shaft tilted forward to fly level at high speeds. Early designs put engines up-front like those of the S-61, but Tobias wanted a sleeker profile. Category A posed a bigger engine dilemma. According to Bill Paul, "Allison had the only engine that had the fuel consumption and the power that barely made it." Customers and engineers needed convincing, and Sikorsky flew tests with a three-bladed CH-53D to substantiate the smaller S-76. "We had actual data that had the rotor loading equivalent to what the S-76 would have on a percentage basis and proved we could meet that Cat. A with the S-76."

The Spirit first flew on March 13, 1977, and the first production example went to Air Logistics Division of Offshore Logistics, Inc., in February 1979. *Sikorsky News* in March 1980 reported 47 deliveries against orders for 327 helicopters. Air Logistics was serving a rig 125 nm offshore with three flights a day. Bristow Helicopters became the first European operator of the S-76 in 1980, and Gerald Tobias told an Offshore Technology Conference in Houston, "The Spirit is proving itself to be the most efficient helicopter of its class

Tests with a three-bladed CH-53D proved Category A performance with the smaller, four-bladed S-76.

for offshore oil work around the world." Okanagan Helicopters put four S-76As in Australia for ESSO and the Woodside consortium and two in the Gulf of Thailand for Union Oil. Bill Paul became Sikorsky president and chief operating officer in 1983.

The Spirit name was soon abandoned, and falling oil prices in the 1980s slashed S-76 sales. The S-76A had nevertheless grown popular as an executive transport. Better performance with the powerful Pratt & Whitney PT6 engine led to the S-76B flown on June 22, 1984. Successive S-76 versions introduced progressively more modern Turbomeca Arriel engines. The first S-76C flew on May 18, 1990 and obtained FAA certification on March 15, 1991.

The S-76C+ was delivered to Norsk Helikopter in July 1996 to fly over the North Sea. Brazilian offshore operator Lider Aviacao bought four S76-C++ in 2006. The 11,875 lb S-76D with Pratt & Whitney PW210S engines, composite rotor blades, and other improvements was delivered to Bristow in 2014. Late that year, Titan Helicopter Group accepted the first S-76D in Africa. In 2019, National Helicopter Services Ltd. deployed two S-76Ds in Guyana for offshore support.

S-76s operators such as Norsk Helikopter Service quickly established the helicopter in offshore oil service.

Sikorsky moved S-76 production from Bridgeport, Connecticut to West Palm Beach, Florida, to Straford, Connecticut and finally to Coatesville, Pennsylvania where engines, systems, and blades were integrated with airframes made by Aero Vodochody in the Czech Republic. A few S-76D air-

Malaysia Helicopter Service operated the S-76C in the South China Sea.

frames were made in China before Lockheed Martin closed the Coatesville Heliplex in 2021. Precise numbers are unavailable, but about 60 to 70% of the 850-odd S-76s served oil operators. The last S-76D from Coatesville was delivered to an offshore oil operator in India.

Global Helicopter

Commercial sales of the rugged S-70 Black Hawk were confined to foreign military and government operators, but Sikorsky studied a 29-seat

S-70C-29 airliner in the 1980s. Studies of a big-cabin S-92 based on Black Hawk dynamics began again in mid-1990, and a mockup of a 19 seat S-92 Helibus was unveiled in 1992 in Las Vegas at the Helicopter Association International

China Southern Airlines flew the Arriel-engined S-76C to oil rigs in the South China Sea.

show. International safety regulations and industry requirements ultimately made the production helicopter a totally new design with flaw-tolerant structures, a new four-stage planetary gearbox, wide-chord composite main rotor blades, GE CT7-8 engines, and four-screen integrated 'glass cockpit.'

The S-92 first flew on December 23, 1998 at West Palm Beach. In December 2002, the 26,500 lb S-92 became the first helicopter certified under harmonized FAA/JAA Part 29 regulations for transport rotorcraft. PHI took first customer delivery in 2004. In 2007, PHI initiated talks with Sikorsky about easy-to-use automatic approach technology. S-92 operators today can use rig approach software to bring the helicopter to a hover a quarter mile from the oil platform hands-off. Automation enabled oil operators to fly more paying sorties and gave crews better situational awareness in reduced visibility. S-92A project pilot Ron Doepner told Vertiflite magazine in 2009, "They will get more people to oil rigs with the system than without, but to my mind, the real payoff is the safety."

Artists concept of the S-70C-29 Advanced Commercial Helicopter, a stretch version of the S-70 Black Hawk

With airframe modules supplied by international partners working in a digital design environment, early S-92s were assembled in Stratford, Connecticut and completed at Keystone Helicopters in Coatesville, Pennsylvania.

Canadian offshore oil operator Cougar Helicopters received its first S-92 in 2005. Sikorsky acquired Keystone that year, and the Coatesville Heliplex assumed complete S-92 assembly and integration responsibilities by the winter of 2009–2010. Líder Aviation in Brazil ordered three S-92s in 2012 for offshore operations. Bond Helicopters Australia received the first of four S-92As in 2015 for offshore services in the Australasian region. In 2017, Coatesville delivered an S-92A to Bristow Norway readily convertible from offshore transport to Search And Rescue configuration for operations in the Barents Sea.

In 2017, Sikorsky, PHI and Outerlink Global Solutions introduced S-92 real-time health-and-usage monitoring with helicopter health data transmitted via satellite. The 300th production S-92 was delivered to Era Group in January 2018 for offshore operations in the Gulf of Mexico. In 2021, oil and gas operators accounted for 86% of all S-92 flight hours. Sikorsky President

The S-92 flew to oil rigs during a mid-east sales tour.

Paul Lemmo told the HAI Heli-Expo in February 2022 "Our average offshore S-92 is flown 800 hours per year. There are many aircraft out there in fleets that never reach 800 hours per year. In fact, many of ours are flying well over 1,000, but the average customer is flying 800 hours with 93% availability."

In 2021, the U.S. Department of the Interior Bureau of Ocean Energy Management counted 75 "medium" and 110 "heavy" Sikorsky helicopters operating in the Gulf of Mexico alone. Oil industry helicopter demand has always fluctuated with crude price-per-barrel. Sikorsky relocated S-92 production to West Palm Beach and with plans for improved S-92A+ and S-92B helicopters, the company continues to cultivate the cyclic but essential oil helicopter market.

PHI in 2022 won a new contract to fly S-92s for Australia's largest natural gas producer.

Chapter 6
Engineering Evolution and Epilog

S-95 CH-53K

S-103 Defiant

Engineering Evolution and Epilog

"All the history of engineering and scientific development of the last century definitely indicates that the impossible of today becomes probable tomorrow and is being accomplished the day after tomorrow. Therefore, the development of this nature does not seem impossible even though at present we may not know how it can be done."

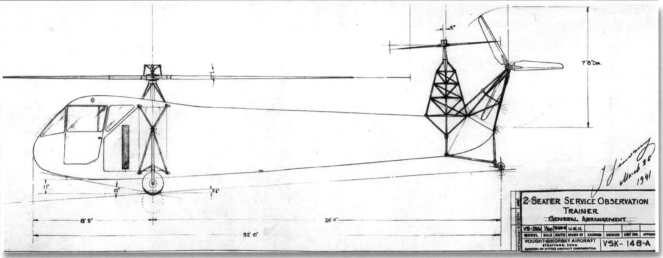

A 1941 general arrangement drawing of a conceptual observation trainer. Note Igor Sikorsky's approval signature on right.

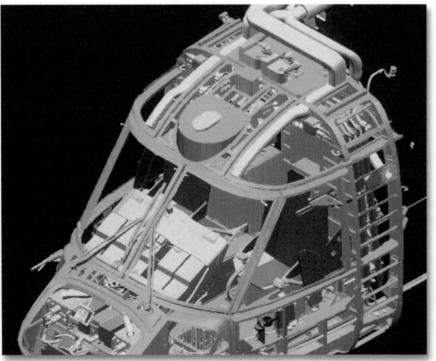

A three-dimensional Computer Aided Design (CAD) model of the S-92 cockpit.

Sikorsky Aero Engineering Corp. was founded in 1923 by an intuitive engineer. Nearly a hundred years later, Sikorsky Aircraft, now a Lockheed Martin Company, has about 3,500 engineers using powerful computer tools to advance the helicopter. Of his father's engineering legacy, Sergei Sikorsky said, "At a minimum, you could say that Igor Sikorsky gave the world a number of interesting inventions. First of all, he pioneered the multi-engine aircraft, which many experts around 1910 said would never be done. Then, during the 1930s, my father gave the world some pretty high-performance flying boats and the Clipper amphibians. They allowed Pan American

The S-21 'Grand' flown in 1913 was the first successful multi-engined airplane.

Airways almost single-handedly to establish long-range over-water flights with their Pacific and Atlantic passenger services." Soon after, Sikorsky engineering refined the helicopter. "In Dad's words, 'The jet may have made the world smaller. The helicopter made it bigger by allowing mankind to live and work in areas that would have been inaccessible by any other vehicle.'"

Igor Sikorsky took a hands-on approach to engineering. "During the period of his flying boats, if there was some sort of problem, Dad would come home, have supper, and then drive back to the factory. Sometimes, he would be alone except maybe for the night watchman -- in those days, single shifts were the rule rather than the exception. He'd come back into the factory, walk around, and mentally engineer whatever problem there was. And often in the middle of the night he would sketch it out, and bingo, he had the solution."

Igor Sikorsky's engineering genius saved the United Aircraft division that carried his name. "When he was finishing up the last of the big flying boats," Sergei recalled, "Dad was beginning to sense a crisis for the future of Sikorsky Aircraft. He would work on the flying boats during the day,

but starting around 1936, he would come home, spend about an hour having a leisurely dinner, take a 15- to 20-minute hike around the back of our property, and then go upstairs into his own private little office and work there sometimes until midnight. It was there that I would see his first sketches of the first helicopters, and Dad was already zeroing in at that time on the configuration of a single main lifting rotor."

Sketches took form in the experimental VS-300. Igor Sikorsky's cousin "Prof" Sikorsky, Alexander "Nick" Nikolsky, and brothers Serge and Michael Gluhareff made up the VS-300 senior engineering staff. Tethered flights in September 1939 began a cycle of helicopter discoveries and engineering changes. Sergei Sikorsky recalled, "I would say the high point, according to my father, was that very first lift-off and two or three weeks later in December 1939 as he realized he had a machine that could fly."

Excessive vibration and other flight test discoveries were addressed by trial-and-error. The VS-300 initially refused to fly forward. Sergei Sikorsky explained, "The problem was they, at that time, underestimated the power of the lead-lag phenomena when flying forward. They had very, very primitive rubber washer-snubbers to control the fore-and-aft movement of the rotor blades. It was shortly after that that they put the hydraulic dampers horizontally so they would dampen the forward and the rearward motion of the rotor blades as the helicopter flew forward. That was one of the major solutions tested on the VS-300."

The S-42 Pan American Clipper flew in 1934 and opened long, over-water passenger routes.

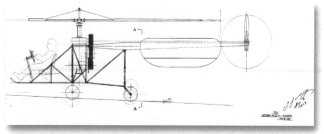

The VS-1 preliminary design – Signed by Igor Sikorsky May 8, 1939, depicted the single main rotor helicopter concept.

With a fly-fix-fly approach, the helicopter became a useful platform. Sergei Sikorsky recalled, "Mechanically, technically, Dad was always impressed by three or four aircraft design factors. One was gross weight of an aircraft versus useful load. I remember several times when I had the privilege of escorting him at the Paris Airshow or at Farnborough. He would always look up and kind of study an aircraft and make a couple of very incisive comments. For instance, he'd say, "This aircraft is going to be a bear to fly – too much of the airplane is forward and not enough in back." Another measure he would always make was the speed range of an aircraft – your stalling speed versus your cruising speed. It was the band between beginning to fly and the top speed could you stretch out of that aircraft with reasonable fuel consumption. If the aircraft could reach a three-to-one ratio, he'd say that was a well-designed, well-balanced aircraft."

Sikorsky engineering overcame the unique challenges of the helicopter. "One of the times I saw my father truly relaxed and smiling was in Stratford, just after he'd established the world's endurance record for helicopters with the VS-300. There was a smattering of press people there and maybe 200 people from the factory who had come out in back of the old Sikorsky plant on the Housatonic River. I remember very clearly his flying or hovering the helicopter. When the sign was held up by one of the Sikorsky mechanics saying that Dad had just established a new record, you could see the smile on his face even though he was hovering maybe 50 feet away."

Building Teams

The VS-300 spawned production helicopters designed by a growing team of Sikorsky Aircraft engineers. In a September 2019 article for the Vertical Flight Society magazine Vertiflite, Ray Leoni recalled, "When I started work in June 1951 there were only 200 people in the engineering department. That number grew to 2,500 when I retired

The VS-300 established a world endurance record of 1 hour 20 min in May 1941 and showed the Sikorsky helicopter was a viable product.

41 years later. In 1951, the company was starting to rebuild its staff and production capacity as a result of the war in Korea and the military's need for more transport helicopters. For Sikorsky, that meant ramping up the S-55 and S-58 production lines eventually setting annual production records that peaked in 1957."

Leoni recalled, "About half of the engineering department consisted of Russian immigrants who left Russia at the time of the revolution. Many sought out Igor Sikorsky when he created his own company on Long Island. They were very

Igor Sikorsky, 'flies' the VS-300 simulator in 1938 surrounded by engineers Michael Buivid, Boris Labensky and Michael Gluhareff.

Increased production was matched by a growing engineering force at the Stratford North Main Street facility.

motivated and dedicated people and very willing to help young guys who knew nothing about helicopters." Leoni continued, "There were no helicopter textbooks. There were no courses about helicopters in college and no company-sponsored educational opportunities at that time. To make learning even harder, there were no design manuals, no lessons-learned reports, and in fact there were very few technical reports. The main source of technical information was in peoples' memories, so you needed to extract information from their heads.

"I recall being in Prof Sikorsky's office asking for help in calculating rotor performance. Professor Sikorsky, as he was called, was Igor Sikorsky's cousin who had become the company's chief aerodynamicist. When I sat down next to his desk, I noticed that his extra-long K&E slide rule was full of burn marks because he often rested his cigarette on the slide rule while doing something else. I asked Prof what he would do if the cursor ended up directly over a burn mark obscuring the answer to his string of calculations. Without hesitating, Prof replied, 'I make-ed verry good guess.'"

Engineering at Sikorsky remained a hands-on activity. Leoni noted, "I remember starting my drafting career when everyone drew with pencil on paper. Some years later we changed from pencil on vellum to ink on Mylar and thought that was fantastic until you needed to make changes to the inked design. Much later, computer-aided design was invented which led not only to much greater engineering productivity but also to much better accuracy plus the ability to have remote design teams work together."

Leoni led design efforts on the Utility Tactical Transport Aircraft System (UTTAS) that would become the S-70 Black Hawk. "When Mr. Sikorsky

Paper and Mylar UTTAS drawings led to a mockup used to demonstrate the integrated maintainability features of the new helicopter.

Sikorsky engineers tested an instrumented S-70 model tested in the United Technologies wind tunnel.

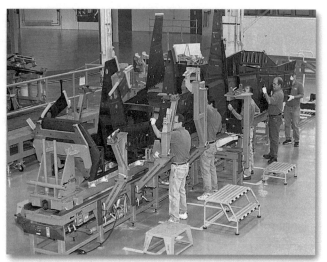

Bridgeport workers assembled Comanche No. 3 in November 2002 with graphic work instructions from CAD data.

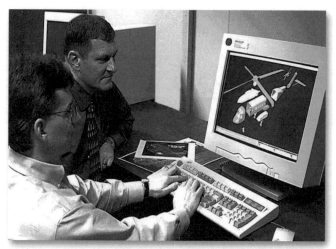

The S-92 had a 3-D electronic mockup shared with international partners.

was told about my design, he came over to my drawing board to see what it was all about. He said in such a nice way: 'Mr. Leoni, this is very ingenious but if we only modified this small part of your design, if we only changed this part, and perhaps this other part that would make it even better.' I ended up changing much of the design, but he made me feel like his changes were of my creation and not his. . . But he did ask me why the main rotor was so close to the top of the fuselage. I explained the need for a compact airframe to meet air transport requirements. Mr. Sikorsky understood the explanation, but he clearly felt uneasy about the main rotor location. He had at that moment pointed out what proved to be a major flaw in our initial UTTAS design that was later corrected by raising the rotor 15 inches. His intuition was right-on."

Digital Tools, Digital Twins

The UTTAS request for proposals was released in 1972, and the Sikorsky solution took shape in a two-dimensional design environment of paper and Mylar. When Sikorsky and Boeing formed their LHX (Light Helicopter Experimental) First Team in 1988 to build the Army's next scout-at-tack helicopter, the RAH-66 Comanche, engineers adopted 3-D digital design and manufacturing plans. The Comanche database

was stored in IGOR, the Stratford electronic drawing vault where it could be updated and shared by industry team members in different locations. Control points on the Comanche assembly line in Bridgeport had computer interfaces for assembly mechanics to call up graphic work instructions. The benefits of the use of digital technology soon became evident. The schedule for the initial assembly of the basic fuselage structure was one and a half weeks. The actual assembly took only a half day. The parts manufactured with digital design fit together so precisely no trim-and-fit was required.

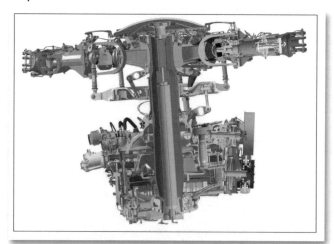

A preliminary CAD drawing shows the CH-53K Quick Change Assembly.

The international S-92 team led by Sikorsky started preliminary design of the big, new transport helicopter in 1991 with two-dimensional CAD sys-

The Black Hawk engineering simulator tests changes in the UH-60M Systems Integration Lab.

tems that replicated traditional drafting but stored drawings in electronic form. As collaborative design progressed, S-92 changes took form in a complete 3-D electronic mockup.

Computer-controlled torque wrenches on the CH-53K line ensure consistent assembly quality.

When the Marine Corps CH-53K began system development and demonstration in 2005, design of the big heavy-lift helicopter relied totally on 3-D tools to create a digital twin of the real aircraft. Digital twins – high fidelity virtual simulations of physical aircraft – promise to keep pace with changes throughout the real helicopter's life and may reveal component wear to manage maintenance and improvements. They are part of digital factory plans to optimize real-world manufacturing.

In July 2022, Lockheed Martin vice president of enterprise business transformation Mike Ambrose discussed Sikorsky's digital engineering

The Raider-X Future Attack Reconnaissance Aircraft leverages model-based systems engineering.

and manufacturing transformation on the Made in America podcast. Ambrose served Sikorsky for 38 years in positions including chief engineer during the early stages of the fast, long-ranged Future Vertical Lift (FVL) initiative. "Now, the title is Enterprise Business Transformation," he explained. "People will often think of that as kind of like the interface between the design and manufacturing. It is a lot more than that. When I talk about supercomputers and high-performance computing, it starts with understanding how the helicopter's going to fly, even before you build it. There are all kinds of aerodynamic, acoustic, and vibration characteristics that we need to model at a conceptual level so we know what to design."

Sikorsky's proposed S-102 Raider-X Future Attack Reconnaissance Aircraft and SB>1 Defiant-X Future Long Range Assault Aircraft leverage the coaxial rigid rotor compound helicopter configuration demonstrated in the 1970s -- the S-69 Advancing Blade Concept. According to Ambrose, "What is different today is our ability to model how this advanced system works, and we've been able to do it with a really high level of accuracy that was unheard of, even maybe 10 years ago, 15 years ago. We do it with very little actual flight test data. If you go back to Igor Sikorsky, obviously, everything was trial-and-er-

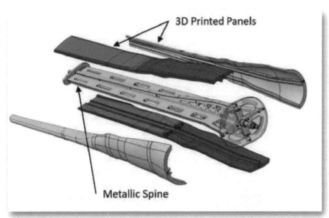

Composite blades for the Raider X were made rapidly with 3-D printed tools.

ror and intuition. We didn't have computers. We didn't have anything to really base it on."

Model-based system engineering today provides a virtual foundation for physical flight test and can shorten historic fly-fix-fly development. "We're able to go and really maximize the data we have, and run through simulations, ultimately using artificial intelligence and machine learning to really predict how this revolutionary technology works. What we've been able to demonstrate over the last couple of years is that we were really good at it, to the point where we could fly the Defiant aircraft from West Palm Beach to Nashville, Tennessee, with a high level of confidence."

The S-92 systems integration lab linked a cockpit simulator to actual flight control hardware to develop control laws and the pilot-vehicle interface.

The Defiant Joint Multirole technology demonstrator was built to test FVL technology and

design tools for the U.S. Army. The company funded-Raider modeled a notional armed scout. "We're able to fly what is basically a helicopter at almost 250 knots," noted Ambrose. "These flights are really validation of the tools and the processes that we had developed using computer simulations enabled by things like high-speed computing. We're able to really model and correlate what we saw in flight test to be able to ensure that the blade design and the shape of the blades and the way we tuned-out the airframe would provide us and the customer with the best vehicle possible."

A digital thread runs all through Defiant and Raider design, development, test, manufacture, and support. Ambrose explained, "The beautiful thing is because we talk about this digital thread, where there's a single source of data, everything's connected. We're simulating how the aircraft's going to perform. At the same time, we've got manufacturing engineers saying, 'Okay, [that's a] great blade shape, but I can't build that. Can you go and tweak this twist so at least I can build it for you?'. As you go through his iterations, we're simulating how we're going to go and fabricate and build it and how we're going to heat [cure] it in ovens."

Ambrose continued, "Once we get that done, we can go and build the tool that you're going to make that blade on. We can 3-D print it. . .[If] we need to go change a part of that -- the way it drops off, or the way it changes its taper. In the past, you would be looking at three to six months to go and make that change. We say, 'Give us a couple of weeks and we're just going to go and additive-manufacture or modify this tool, and you'll have it in three weeks."

The digital thread also pays off in the digital factory where engineering data refines production processes. Ambrose noted, "One of the things that we're always looking at as we go and connect the digital thread. As we look to see how do

The Defiant Joint Multirole Technology Demonstrator proved digital design tools for Future Vertical Lift.

we integrate processes, we also look at how do we integrate suppliers into that. By doing that, we're able to understand what are the processes that we need the supply base to focus on to be able to optimize our designs."

The digital factory already controls torque wrenches and other tools on Sikorsky's CH-53K production line to make assembly faster and quality more consistent. Mike Ambrose said, " Our workers embraced the transformation because it allows them to do their job easier, with better quality, less mistakes. The collaboration with our workforce on how we're able to go figure out these things together is an example of how, I like to think, engineers are pretty smart. We also have some really smart mechanics, who

The 10th H-60 Multi-Year Procurement contract gives Si-korsky solid business building UH-60Ms and HH-60Ms (shown) for the near future. (U.S. Army)

are there every day putting these things together, and they provide that feedback to our manufacturing engineers on how we can do it even better."

Epilog

By 2022, Sikorsky Aircraft, a Lockheed Martin Company, had 13,000 people designing, building, and supporting helicopters worldwide. Established products booked good business for the near fu-ture. The 10th H-60 Multi-Year Production contract was announced in June. Stratford workers were building UH-60M and HH-60M Black Hawks for the U.S. Army and Foreign Military Sales, CH-53K heavy lifters for the U.S. Marine Corps, and HH-60W combat rescue helicopters for the U.S. Air Force. Stratford and Owego teams were ready for new MH-60R Naval Hawk orders from Korea, Greece, and Australia. Sikorsky Mielec in Poland was making Black Hawks for Romania, Poland, and the Philippines. Sikorsky and United Rotorcraft was integrating Firehawks for public service and commercial operators. With 13 heads of state already flying in Sikorsky S-92As, the VH-92 was in full-rate production for the President of the United States.

Sikorsky X2 technologies give the company high-speed compound helicopters such as the Defiant Joint Multirole Technology Demonstrator. (U.S. Army)

Sikorsky leadership around 2004 envisioned future products built around speed, autonomy, and intel-ligence or self-awareness. Sikorsky Innovations stood up in 2009, and the rapid-prototyping

The company-funded S-97 Raider uses X2 technologies to enhance the value of the helicopter with more speed, agility, and payload. (Sikorsky)

The S-76 SARA provided an autonomy testbed to develop Matrix technologies (Sikorsky)

team took the X2 Raider compound helicopter to 250 kt in 2010. The suite of X2 technologies – coaxial composite rotors, integrated auxiliary propulsion, fly-by-wire flight controls, and active vibration sup-pression – was Sikorsky's key to Future Vertical Lift (FVL) offerings. The U.S. Army envisions FVL as a family of scalable rotorcraft to double helicopter speed and range around 2030.

The Army also expects Future Vertical Lift aircraft to fly optionally-manned with some level of flight control autonomy. Supervised autonomy and inte-grated sensors promise to offload human pilots and enhance safety and productivity in difficult conditions. Full autonomy gives warfighters aircraft that can deploy over long distances or fly dangerous missions without on-board crew. The Sikorsky vision is a cockpit readily switched from two pilots to one or none depending on mission.

The Sikorsky Matrix architecture integrates hardware and software for helicopters to execute complex missions close to obstacles. It first flew in 2013 on the S-76 fly-by-wire Sikorsky Autonomy Research Aircraft (SARA). Under the Aircrew Labor In-Cockpit Automation System (ALIAS) program of the Defense Advanced Research Projects Agency, Matrix enabled untrained pilots to fly the helicopter through a tablet computer. The platform-agnostic technology also flew fixed-

wing turbo-props with the same pilot interface. In June 2019, a company-owned UH-60A with Matrix kit and safety pilots flew as an Optionally Piloted Vehicle. In 2022, the ALIAS Black Hawk flew through tactical courses and delivered cargo totally uninhabited.

Sikorsky's third pillar of advanced flight is an intelligent, self-aware aircraft that monitors its own condition to enhance safety and reduce maintenance. The S-92 implemented an Integrated Mechanical Diagnostic System)/Health-and-Usage Monitoring System – IMDS/HUMS – that collects vibration data from bearings, driveshafts, and other dynamic components to cue maintainers and fleet managers. It has been tied to satellite communications to give managers on the ground real-time health monitoring in flight. By 2022, the S-92 fleet report-ed better than 93% availability in high-tempo oil industry operations, in part due to IMDS/HUMS insights.

Matrix autonomy was demonstrated by an uninhabited UH-60A Black Hawk during the ALIAS program for DARPA. (Sikorsky)

At a Vertical Flight Society digital enterprise workshop, Sikorsky Senior Technical Fellow Nick Lappos suggested future aircraft be considered cyber-physical entities: smart systems that integrate analog, digital, physical, and human components with physics and logic. He said, "The aircraft actually depends on the cybernetic as much as spars and rotor blades. . . . When you trust the digital systems that much, you'll get the benefits." Knowing actual engineering limits may make future aircraft lighter and more productive. Connecting physical and data dimensions with world-wide analytics, monitoring and control can also enable systems to "learn" and get better with time.

In 2022, Sikorsky Aircraft added a fourth pillar to the Innovations portfolio. Distributed electric pro-pulsion opens design possibilities for quieter, cheaper vertical flight. Sikorsky plans a hybrid-electric demonstrator that will use fuel to generate electricity and drive new lifting configurations.

The company started by Igor Sikorsky in 1923 gave the world new and better ways to fly. It stands ready today with the technology, tools, and talent to continue Sikorsky's vision.

Sikorsky tools create a digital thread running through CH-53K design, development, manufacture, and sustainment.

Appendix 1
Sikorsky Aircraft History

"The conquest of the air, with its profoundly important influence on the history of humanity will remain one of the outstanding facts of the twentieth century."

S-29 under construction at the first Sikorsky Facility – Utgoff Farm, Long Island, N.Y.

Corporate Facilities

Long Island, N.Y. - 1923

Long Island, N.Y. - 1924

College Point, N.Y. - 1926

Lordship, Stratford, Ct. - 1929

Bridgeport, Ct. - 1943

Stratford, Ct. - 1955

Corporate Identity

SIKORSKY AERO ENGINEERING CORPORATION
250 West 57th Street New York, N.Y.

March 1923 - July 1925

Sikorsky Manufacturing Corporation
College Point, New York

July 1925 - October 1928

October 1928 - July 1929

SIKORSKY AVIATION CORPORATION
DIVISION OF UNITED AIRCRAFT AND TRANSPORT CORP.

July 1929 - January 1935

Sikorsky Aircraft Division of United Aircraft Manufacturing Corporation

January 1935 – July 1936

July 1936 – April 1939

VOUGHT-SIKORSKY AIRCRAFT
DIVISION OF UNITED AIRCRAFT CORPORATION
STRATFORD, CONNECTICUT

April 1939 – December 1942

January 1943 – January 1975

January 1975 – October 2015

November 2015 - present

Management

"From the beginning of 1924 until the latter part of 1926 our factory was located in one of the two old wooden hangars on a corner of Roosevelt Field near Westbury, Long Island. At that time there was no permanent night watchman and those duties were carried out by someone in the organization, once in a while by myself."

Igor Sikorsky

March 1923 to July 1925:
- President and Chief Engineer – Sikorsky Aero Engineering Corp

July 1925 to October 1928:
- President – Sikorsky Manufacturing Corporation

October 1928 to May 1957:
- Vice President and Chief Engineer – Various Sikorsky Divisions of United Aircraft

May 1957 to October 1972:
- Consultant - Sikorsky Aircraft Division of United Aircraft Corporation

Products Introduced / Produced while president
S-29A, S-31, S-32, S-33, S-34, S-35, S-36, S-37, S-38

Arnold Dickinson

October 1928 to August 1929:
- President – Sikorsky Aviation Corp.

August 1929 to April 1930:
- President - Sikorsky Aviation Corporation, a Subsidiary of United Aircraft and Transport Corporation

Products Introduced / Produced:
S-38, S-39

Eugene E. Wilson

April 1930 to December 1931:
- President – Sikorsky Aviation Corporation, a subsidiary of United Aircraft and Transport Corporation

Products Introduced / Produced:
S-40, S-41

Frederick Neilson

October 1928 to December 1931:
- President – Sikorsky Aviation Corp.

July 1935 to December 1935:
- President - Sikorsky Aviation Corporation, a Subsidiary of United Aircraft and Transport Corporation

Products Introduced / Produced:
S-41, S-42, S-43

Renssalaer W. Clark

December 1935 to July 1936:
- General Manager – Sikorsky Aircraft Division of United Aircraft Manufacturing Corporation

July 1936 to April 1939
- General Manager – Sikorsky Aircraft Division of United Aircraft Corporation

April 1939 to April 1939:
- General Manager – Vought-Sikorsky Aircraft Division of United Aircraft Corporation

Products Introduced / Produced:
S-43, S-44 (XPBS-1), F4U Corsair, OS2U-1 Kingfisher

Charles McCarthy

April 1939 to January 1943:
- General Manager – Vought-Sikorsky Aircraft Division of United Aircraft Corporation

January 1943 to February 1943:
- General Manager – Sikorsky Aircraft Division of United Aircraft Corporation

Products Introduced / Produced:
VS-300 (S-46), VS-44, S-47 (R-4), F4U Corsair

J. Reed Miller

February 1943 to October 1943:
- General Manager – Sikorsky Aircraft Division of United Aircraft Corporation

Products Introduced / Produced:
S-47 (R-4), S-48 (R-5), S-49 (R-6)

Bernard L. Whelan

October 1943 to January 1957:
- General Manager – Sikorsky Aircraft Division of United Aircraft Corporation

Products Introduced / Produced:
S-46, S-48, S-49, S-51, S-52, S-53, S-54, S-55, S-56, S-58, S-59

Lee Johnson

January 1957 to September 1968:
- General Manager / President – Sikorsky Aircraft Division of United Aircraft Corporation

Products Introduced / Produced:
S-55, S-56, S-58, S-60, S-61, S-62, S-64, S-65, APBS-1 (patrol boat),

Wes Kuhrt

September 1968 to January 1974:
- President – Sikorsky Aircraft Division of United Aircraft Corporation

Products Introduced / Produced:
S-58, S-61, S-62, S-64, S-65, S-67, S-69

Gerald Tobias

January 1974 to May 1975:
- President – Sikorsky Aircraft Division of United Aircraft Corporation
May 1975 to January 1981:
- President - Sikorsky Aircraft Corporation, a subsidiary of United Technologies Corporation

Products Introduced / Produced:
S-61, S-64, S-65, S-70A, S-72 (Rotor System Research Aircraft (RSRA), S-76, S-65 (CH-53E)

Robert Daniell

January 1981 to September 1983:
- President – Sikorsky Aircraft Corporation, a subsidiary of United Technologies Corporation

Products Introduced / Produced:
S-61, S-65, S-70A, S-70B, S-76

William Paul

September 1983 to October 1985:
- President – Sikorsky Aircraft Corporation, a subsidiary of United Technologies Corporation

Products Introduced / Produced:
S-65, S-70A, S-70B, S-75 (Advanced Composite Airframe Project), S-76

Bob Zincone

October 1985 to July 1987:
- President – Sikorsky Aircraft Corporation, a subsidiary of United Technologies Corporation

Products Introduced / Produced:
S-65, S-70A, S-70B, S-76

Gene Buckley

July 1987 to August 1999:
- President – Sikorsky Aircraft Corporation, a subsidiary of United Technologies Corporation

Products Introduced / Produced:
S-65, S-70A, S-70B, S-76, S-80, S-92, RAH-66, Cypher

Dean Borgman

August 1999 to June 2003:
- President – Sikorsky Aircraft Corporation, a subsidiary of United Technologies Corporation

Products Introduced / Produced:
S-65, S-70A, S-70B, S-76, RAH-66

Steve Finger

June 2003 to March 2006:
- President – Sikorsky Aircraft Corporation, a subsidiary of United Technologies Corporation

Products Introduced / Produced:
S-70A, S-70B, S-76, S-92, RAH-66

Jeff Pino

March 2006 to July 2012:
- President – Sikorsky Aircraft Corporation, a subsidiary of United Technologies Corporation

Products Introduced / Produced:
S-70A, S-70B, S-76, S-92, S-94 (X-2)

Mick Maurer

July 2012 to April 2015:
- President – Sikorsky Aircraft Corporation, a subsidiary of United Technologies Corporation

Products Introduced / Produced:
S-65, S-70A, S-70B, S-76

Bob Leduc

April 2015 – November 2015:
  • President – Sikorsky Aircraft Corporation, a subsidiary of United Technologies Corporation

Products Introduced / Produced:
S-70A, S-70B, S-76, S-92, S-95 (CH-53K) & S-97 Raider (First Flights in 2015)

Dan Schultz

November 2015 – December 2020:
  • President – Sikorsky Aircraft Corporation, a Lockheed Martin Company

Products Introduced / Produced:
S-70A, S-70B, S-76, S-92, S-95, S-97, S-103

Paul Lemmo

December 2020 - Present
  • President – Sikorsky Aircraft Corporation, a Lockheed Martin Company

Products Introduced / Produced:
S-70A, S-70B, S-76, S-92, S-95, S-102

Flight Heritage

"How many more new types of flying ships under the Winged-S will rise in the air after the S-44 Flying Dreadnaught and the R-6 Helicopter cannot be surmised. Whatever their number, if any, they will in due time be replaced and by far surpassed by other larger, faster, and more luxurious flying vessels that will cross continents and oceans, tropical and polar regions, with remarkable efficiency, comfort and regularity."

S-29A

First flight: May 4, 1924. The S-29A was the first airplane Igor Sikorsky built in the United States. With two 400 hp engines, it could carry 14 passengers and had a cruising speed of 100 mph. One S-29A was built.

First flight: September 13, 1925. The S-31 with a single 200 hp engine was designed for the U.S. Airmail Service. It could also be used for observation/bomber, photographic platform, or four-passenger transport. One was built.

S-31

S-32

First flight: December 6, 1925. The S-32 with a 400 hp engine was used to ferry supplies for the Andean National Corporation, a subsidiary of The Standard Oil Company. One was built.

S-33

First flight: December 1925. The S-33, with a 60 hp engine, was a two-place wooden biplane using the Gluhareff-Sikorsky wing design for enhanced control, performance, and safety. Two were built.

First flight: 1926. The S-34 was Igor Sikorsky's first amphibious aircraft. It could carry six passengers and was powered by two 200 hp engines, but performance was disappointing. One was built.

First flight: August 20, 1926. The S-35 with three 425 hp engines was built to compete for the Orteig Prize for the first non-stop flight between New York and Paris. Only one was built.

First flight: August 1927. The S-37 was powered by two 500 hp engines and had a 100-foot wingspan. It had a range of 4,000 miles at 120 mph and provided the first commercial air transport between Buenos Aires, Argentina and Santiago, Chile, regularly crossing the 19,000 ft Andes Mountains. Two were built.

First flight: August 20, 1927. The S-36 with twin 200 hp engines was a modified and enlarged version of the S-34 but had an enclosed cabin for passengers and cargo. Six were built.

First flight: June 25, 1928. The S-38 twin-engine amphibian could carry eight passengers and two crew. Its commercial success allowed Igor Sikorsky to reorganize as Sikorsky Aviation Corporation and relocate to Stratford, Connecticut. One hundred and eleven were built.

First flight: December 24, 1929. The S-39 was a 300 hp single-engine derivative of the S-38. It carried five passengers for executive and transport service. Twenty-one were built.

S-41

First flight: May 4, 1930. The S-41, with two 575 hp engines was an amphibian much like the S-38. It had a longer wing span and could carry 14 passengers. Seven were built.

First flight: August 7, 1931. The S-40, 'The American Clipper,' was a large amphibian with four 575 hp engines. It carried 40 passengers, weighed over 34,000 pounds, and was designed for Pan American Airways. Three were built.

S-40

S-42

First flight: March 30, 1934. The S-42, with four 750 hp engines was the largest aircraft of its day. It could carry 35 passengers. In service with Pan American Airways, it established 10 world records, eight in one day. Ten were built.

First flight: June 1, 1935. The S-43 - The Baby Clipper - was a very successful twin-750 hp engine airplane. It could carry 15 passengers and was used in military and commercial service. Fifty-three were built.

S-43

S-44

First flight: August 13, 1937. The S-44 flying boat (U.S. Navy designation XPBS-1 shown) was a long-range patrol bomber with four 900 hp engines. One was built. Three commercial configurations, designated as the VS-44, were also built.

S-46

First flight: September 14, 1939. Also known as the VS-300, this testbed, which took only nine months to design and build, pioneered and refined the world's most common helicopter configuration.

S-47

First flight: January 14, 1942. The two-seat S-47 was Sikorsky's first production helicopter and the only helicopter to see service with the allies during WW II. It was given the military designation R-4 by the U.S. Army "shown" and HNS-1 by the Navy and Coast Guard. 131 were built.

S-48

First flight: August 18, 1943. The S-48 (R-5A shown) with a 450 hp engine, was a two-seat tandem cockpit aircraft with the observer in front and pilot in the rear. It could carry two casualty litters, one on each side of the cabin. 65 were built.

S-49

First flight: October 15, 1943. The S-49 (Army R-6 shown) with a 240 hp engine was a two-seat, single-engine helicopter. It was produced in quantity under license by Nash-Kelvinator. 233 were built.

First flight: February 16, 1946. The S-51, a derivative of the R-5, had seating for three passengers and a pilot. This helicopter performed the first naval and civil rescues. 379 were built.

S-51

S-52

First flight: February 12, 1947. The S-52, a two-seat civil helicopter, was the first production helicopter to use all-metal rotor blades. This aircraft established both speed and altitude records and was the first to perform an inside loop. A military version for the U.S. Marines was also constructed. 95 were built.

First flight: September 22, 1947. The S-53 was a three-place helicopter requested by the U.S. Navy (Navy XHJS-1 shown) for light utility and search and rescue missions. Three prototypes were built.

S-53

S-54

First flight: December 20, 1948. The S-54, utilizing components of an S-47, was converted into a 'sesqui-tandem' rotor helicopter. It had an observer's seat to the rear of the main rotor. The aircraft had a total flight time of 4 hrs. 25 mins. One was built.

First flight: November 10, 1949. The S-55 (Army UH-19 shown) could carry up to ten passengers. It was used by all the U.S. armed services and flew the first commercial scheduled passenger services. A total of 1,758 S-55's were built by Sikorsky and licensees.

S-55

S-56

First flight: December 18, 1953. The twin-engine S-56 (Marine HRS2-1 shown) was designed as an assault transport for the U.S. Marine Corps. It had a five-bladed rotor, retractable landing gear, and clam shell nose doors. The U.S. Army also purchased this helicopter. 156 were built.

First flight: March 8, 1954. The S-58 (Coast Guard HH-34 shown) was a single engine utility helicopter flown by all the U.S. armed services, commercial operators, and many foreign governments. It could carry up to 16 passengers. 2,260 were built by Sikorsky and licensees.

S-58

S-59

First flight: June 11, 1954. The S-59 was the first Sikorsky helicopter with a gas turbine engine. It had four main rotor blades and retractable landing gear. It broke both speed and altitude records. Three were built.

First flight: May 14, 1958. The S-62 had a single turboshaft engine and amphibious boat hull for the U.S. Coast Guard and commercial operators. It could carry 12 passengers or six casualty litters. 170 were built.

S-62

S-61

First flight: March 11, 1959. The twin turboshaft S-61 (Navy H-3 shown) was developed for anti-submarine warfare and was designed with an amphibian hull similar to Sikorsky's fixed-wing flying boats. 1,473 S-61s were built by Sikorsky and licensees.

First flight: March 25, 1959. The twin-engine S-60 demonstrated Igor Sikorsky's dream for a versatile heavy lift helicopter. The 'Flying Crane' was capable of lifting cargo on a pallet and personnel in a detachable pod. One was built.

S-60

S-61

First flight: December 6, 1960. The S-61L is a lengthened baseline S-61 fuselage for commercial passenger operations, carrying 28 passengers and 2 crew. It was the first twin turbine helicopter to be certified for scheduled passenger service.

First flight: May 9, 1962. The twin-turboshaft S-64 Sky-crane (Army CH-54B shown) was designed for heavy lift. It had a pod to carry troops or a field hospital and was used to recover downed aircraft and off-load cargo ships. Commercial versions fight forest fires and support construction and logging. 99 were built.

S-64

S-61

First flight: January 19, 1963. The S-61R variant (Coast Guard HH-3F shown) of the S-61 was outfitted with a rear ramp, a nose wheel, an auxiliary power plant and airfoil shaped sponsons that house retractable landing gear. This aircraft was utilized by both the U.S. Air Force and Coast Guard.

First flight: October 14, 1964. The twin turboshaft S-65 or CH-53A Sea Stallion (shown) was the U.S. Marine Corps assault transport. It could carry 58 passengers or 24 medical litters and could operate in all weather conditions. Versions were adopted by the U.S Navy for mine-sweeping and the U.S. Air Force for combat rescue. 522 were built.

S-65

S-67

First flight: August 20, 1970. The S-67 Blackhawk was developed with company funds as a high-speed gunship that could carry missiles or rockets. It set a world speed record of over 220 knots. One was built.

First flight: July 26, 1973. The twin turbine ABC (Advancing Blade Concept) demonstrator had two counterrotating three-bladed coaxial rigid rotors. The S-69 (Army XH-59A) was later equipped with two P&W J-60 auxiliary turbojets and was capable of speeds over 260 kts. 2 were built.

S-69

S-70

First flight: October 17, 1974. The S-70 prototype (U.S. Army YUH-60A) was Sikorsky's response to the U.S. Army Utility Tactical Transport Aircraft System (UTTAS) requirement. After extensive testing by the Army, Sikorsky won the contract to produce the UH-60 Black Hawk. Versions equip all the U.S. Army services and many international air forces. Over 5,000 S-70s have been built to date.

First flight: October 12, 1976. The S-72 (Rotor System Research Aircraft) was designed in conjunction with NASA and the U.S. Army as a flying testbed for experimental rotors. Two were built.

S-72

S-76

First flight: March 13, 1977. The twin turbine S-76A carries 12 passengers in oil industry, executive transport, and aeromedical operations. More than 600 were delivered in various configurations.

First flight: October 17, 1978. The production S-70 was designated as the UH-60A by the U.S. Army and replaced its fleet of UH-1 helicopters.

S-70

S-70

First flight: December 12, 1979. The S-70B SEAHAWK (shown) was the naval version of the S-70 BLACK HAWK. Its primary mission was anti-submarine warfare, but Seahawks also flew anti-surface warfare, search and rescue, and logistical support missions. 181 were built. Derivatives equipped foreign navies.

First flight: March 1, 1983. A major upgrade to the original S-65 added a third engine and a seven-bladed main rotor to lift 16 tons. This Marine Corps CH-53E (shown) heavy cargo helicopter spun-off the Navy MH-53E minesweeper. 227 were built.

S-65

S-76

First flight: June 22, 1984. The S-76B version of the S-76A designed for hot/high conditions and improved single-engine performance.

First flight: July 27, 1984. The S-75 ACAP (Advanced Composite Airframe Program) was designed to prove the feasibility that fiber-reinforced composite materials could significantly reduce airframe cost and weight. One was built.

S-75

S-70

First flight: March 19, 1987. The SH-60F (shown) was a derivative of the S-70 SH-60B Seahawk with dipping sonar. It served as the U.S. Navy's carrier battle groups primary antisubmarine warfare (ASW) aircraft. 75 built.

S-70

First flight: October 6, 1987. The VH-60N, a modified UH-60L helicopter, is utilized by the U.S. Marine Corps to support their Presidential transport mission. Nine were built.

S-70

First flight: March 22, 1988. The S-70A (Army UH-60L shown) BLACK HAWK was the first major upgrade of the S-70 (UH-60A) BLACK HAWK, with upgraded engines and an improved durability main gearbox, allowing the aircraft to perform at heavier gross weights. Succeeded in production by the S-70M (UH-60M).

First delivery: January 13, 1989. Export version of the MH-53E (S-65 series). Primary missions are airborne mine countermeasures and heavy lift duties. Eleven were built.

S-80

S-76

First flight: May 18, 1990. The S-76C combined the S-76B airframe with higher horsepower (Turbomela Arriel) engines. The S76C series also included the S-76C+ and S-76C++ versions, each with technically advanced engines.

First flight: January 4, 1996. The RAH-66 (Comanche) was a fast, agile, low observable helicopter with tandem seats and two turboshaft engines. The advanced scout/attack helicopter was built by Sikorsky and Boeing for the U.S. Army. Two prototypes were built.

RAH-66

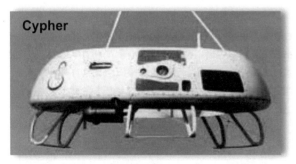

Cypher

First flight: September 1, 1997. An Unmanned Aerial Vehicle (UAV). This proof of concept demonstrator utilized a co-axial shrouded rotor system envisioned to perform search, urban or anti submarine surveillance missions without jeopardizing human life. Two were built.

First flight: December 23, 1998. The S-92 is a 19 passenger, twin turbine civil transport designed for offshore oil, executive transport, and search and rescue operations. The flaw and damage-tolerant design won the Collier Trophy for 2002. Currently in production.

S-92

S-70

First flight: January 27, 2000. Based on the UH-60L BLACK HAWK the U.S. Navy MH-60S (shown) utilizes the engines, drivetrain and rotors of the Seahawk to fly cargo, mine countermeasures, and special warfare support missions. 256 built.

First flight: March 19, 2001. Derivative of the SH-60B, the MH-60R SEAHAWK (shown) is deployed with the U.S. Navy as the primary anti-submarine anti-surface weapon system. Designed to operate from frigates, destroyers, cruisers and aircraft carriers. 275 delivered; international production continues.

S-70

S-70

First flight: September 17, 2003. An extensive modernization of the BLACK HAWK, the UH-60M has composite main rotor blades, an integrated digital cockpit, and machined aerostructures for improved performance, digital connectivity, and reduced operating costs. More than 1,200 delivered.

S-94

First flight: August 27, 2008. The Sikorsky-funded S-94 or X2 high speed technology demonstrator integrated the advancing blade concept with structural, propulsion, and flight control advances. One was built.

S-92

First flight: November 15, 2008. The militarized S-92 entered service as the CH-148 Cyclone for the Canadian Maritime Helicopter Program. Missions (shown) include anti-submarine warfare (ASW), search and rescue and utility missions. 28 aircraft were delivered.

S-76

First flight: February 7, 2009. The S-76D is an advanced derivative of the S-76C with improved performance, reduced external noise, reduced vibration and provided all-weather capability with a blade de-ice system.

First flight: May 22, 2015. The S-97 Raider compound helicopter is a company funded demonstrator integrating the rigid rotor advancing blade concept with fly-by-wire flight controls, integrated tail thruster, and active vibration control to achieve unprecedented speed and agility. Two were built.

First flight: October 27, 2015. The S-95 (Marine Corps CH-53K King Stallion shown) was designed to replace the CH-53E and entered development flight test in October 2015. Currently in production.

First flight: July 28, 2017. Derived from the commercial S-92, the VH-92A Presidential helicopter has a 14-passenger executive interior, upgraded engines and other specialized equipment for the U.S. Marine Corps to support the Presidential transport mission. Currently in production.

First flight: March 21, 2019. The Sikorsky-Boeing SB>1 Defiant Joint Multirole Technology Demonstrator is a high-speed compound helicopter to help the U.S. DoD make informed decisions for Future Vertical Lift (FVL). One was built.

First flight: May 17, 2019. A derivative of the U.S. Army UH-60M, the Air Force HH-60W replaces the HH-60G combat search and rescue helicopters.

Sikorsky / Department of Defense Aircraft Designations
Cross Reference Guide

Sikorsky S-Series	Popular Names	DoD Aircraft Designations				
		Navy	Air Force	Coast Guard	Army	Marines
S-43		JRS-1				JRS-1
S-44	Flying Dreadnought	XPBS-1				
S-47	Hoverfly	HNS-1		HNS-1	R-4	
S-48	Dragonfly				R-5	
S-49	Hoverfly II			HOS-1	R-6	
S-51		HO2S-1G HO3S-1G HO3S-2	H-5F,G HF-5H	HO2S-1G HO3S-1G		HO3S-1G
S-52		HO5S-1	YH-18A	HO5S-1	YH-18, B	HO5S-1
S-53		XHJS-1				
S-55	Chickasaw Whirlwind	HO4S-1,2 UH-19F	YH-19 UH-19A,B HH-19B	HO4S-3G HH-19G	UH-19C,D	HRS-1,2 CH-19E
S-56	Mojave	HR2S-1W			CH-37A,B	HR2S-1 CH-37C
S-58	Choctaw Seabat Seahorse	SH-34G,H,J UH-34G,J HSS-1		HH-34F	CH-34A,C VH-34C	UH-34D,E VH-34D
S-59					YH-39	
S-61	Sea King Sea Knight Jolly Green GiantPelican	HSS-2 CH-3B HH-3A SH-3A,D,G, H,J UH-3A,H VH-3A	CH-3C,E HH-3E	HH-3F		
S-62	Sea Guard			HH-52A		

Sikorsky / Department of Defense Aircraft Designations
Cross Reference Guide

Sikorsky S-Series	Popular Names	DoD Aircraft Designations				
		Navy	Air Force	Coast Guard	Army	Marines
S-64	Tarhe				CH-54A,B	
S-65	Sea Stallion Super Stallion Pave Low Pave Low III, IV Super Jolly Green Giant Sea Dragon	CH-53A,D,E MH-53E RH-53D	MH-53J,M TH-53A			CH-53A,D,E
S-69	ABC (Advancing Blade Concept)				XH-59	
S-70	Pave Hawk Seahawk Black Hawk Quick Fix Jayhawk White Hawk, Jolly Green II, Naval Hawk, Ocean Hawk, Knighthawk	CH-60S HH-60H,J MH-60R,S SH-60B,R,F YMH-60R	HH-60G,W,U MH-60G UH-60A	HH-60J MH-60T HH-60T	YUH-60A EH-60A,B HH-60L,M UH-60A,C,L,M,Q,V MH-60A,K,L	VH-60D,N
S-92	Helibus					VH-92A
S-95	King Stallion					CH-53K
None	Comanche				RAH-66	

DoD Mission Identifiers

CH	Cargo		PB	Patrol Bomber
EH	Special electronic installation		R	Rotor
HC	Helicopter - Crane		RAH	Reconnaissance Attack
HH	Search and Rescue/Medevac		RH	Reconnaissance
HJ	Helicopter - Utility		SH	Anti-submarine warfare
HN	Helicopter - Training		TH	Trainer
HO	Helicopter - Observation		UH	Utility
HR	Helicopter - Transport		VH	Staff
HS	Helicopter - anti-Submarine		X	Experimental
JR	Utility Transport		Y	Prototype
MH	Multi-mission			

Appendix 2
Igor Sikorsky

"The United States seemed to me the only place which offered a real opportunity in what was then a rather precarious profession...There was hope of a great development of air travel, because of the vast geographic space which made the country so favorable for aviation, and also there was the recognition of individual freedom and initiative, which meant so much to me."

Patents, Office and Presentation to Employees

Progress in Patents

Igor Sikorsky's Inventive Genius Produced 66 Patents Awarded In America

Igor Sikorsky was granted 66 U.S. patents. Not all of his ideas attained production, but all proposed fixed- and rotary-wing innovations to make aircraft progressively more capable. The examples that follow explain select ideas, and the original patent drawings show the scope of Sikorsky's inventive genius.

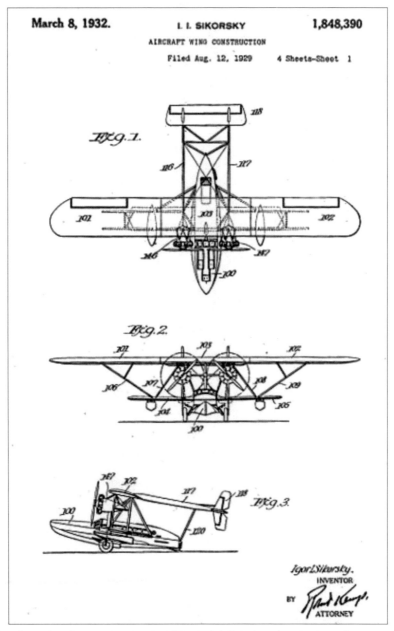

#1,848,390, Aircraft Wing Construction, March 8, 1932 - Igor patents his method mounting the engines on an amphibian by hanging them from the upper wing and goes through the structural method of doing so.

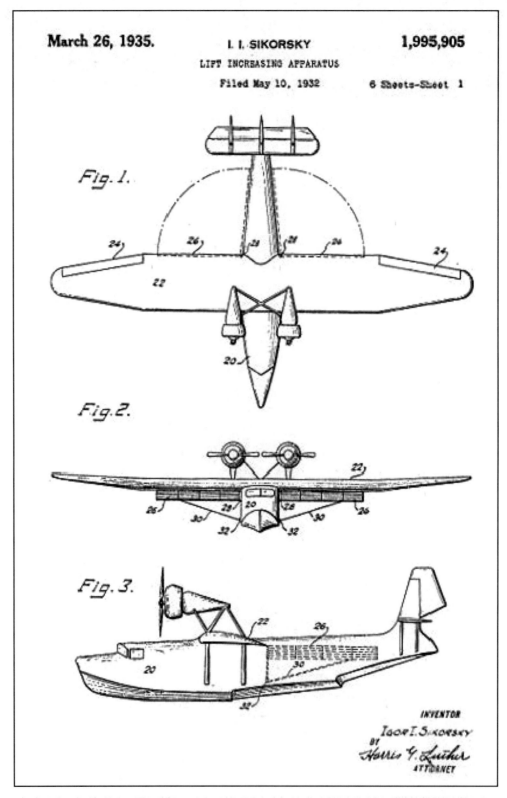

March 26, 1935.   I. I. SIKORSKY   1,995,905

LIFT INCREASING APPARATUS

Filed May 10, 1932   6 Sheets-Sheet 1

Fig. 1.

Fig. 2.

Fig. 3.

INVENTOR

Igor I. Sikorsky

BY Harris G. Luther

ATTORNEY

#1,995,905, Lift Increasing Apparatus, March 26, 1935 - Igor patents methods of increasing the lift of an airfoil at low speeds, including the use of "a plurality of superimposed vanes positioned adjacent and under the rear part of the airfoil." Some of these swiveled 90 degrees to be employed. Others were sensitive to aircraft speed and were automatically retracted at higher speed.

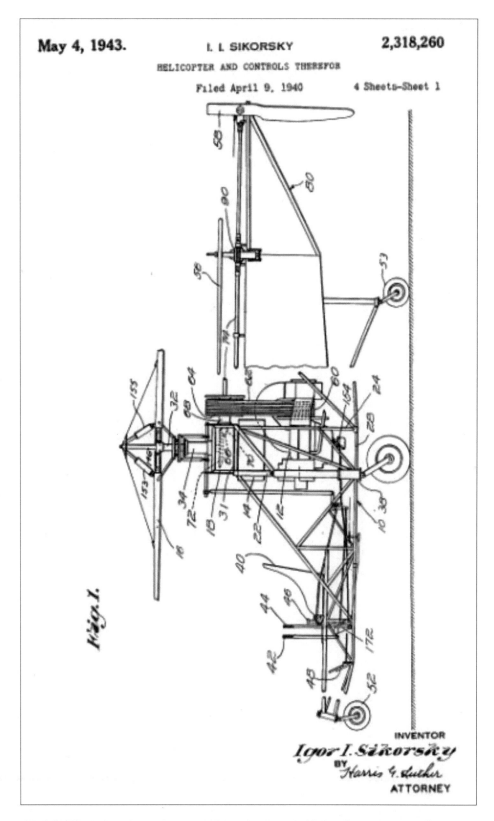

May 4, 1943.   I. I. SIKORSKY   2,318,260

HELICOPTER AND CONTROLS THEREFOR

Filed April 9, 1940   4 Sheets-Sheet 1

Fig. 1.

INVENTOR

Igor I. Sikorsky

BY

Harris G. Luther

ATTORNEY

#2,318,260, Helicopter and Controls Therefor, May 4, 1943 - Covers Igor's early ideas for control of the single rotor helicopter, including three tail rotors - one for anti-torque and yaw control, and two mounted horizontally to control aircraft pitch and roll. Also patents tail rotor pitch control connected with main rotor pitch control.

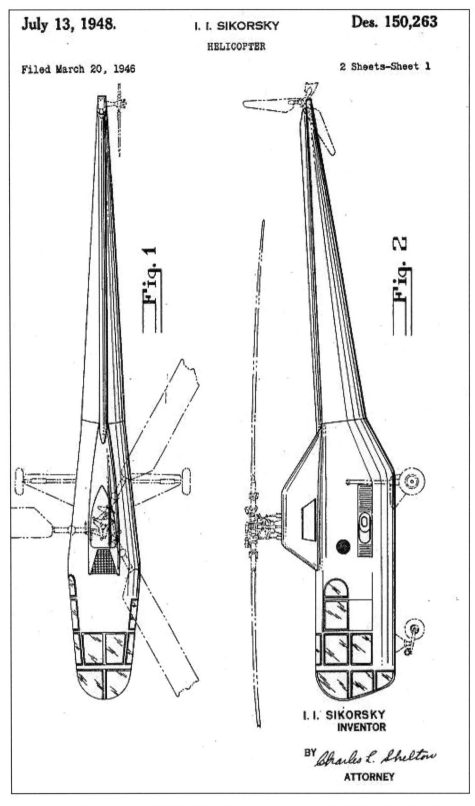

July 13, 1948.

I. I. SIKORSKY

Des. 150,263

HELICOPTER

Filed March 20, 1946

2 Sheets—Sheet 1

Fig. 1

Fig. 2

I. I. SIKORSKY
INVENTOR

BY *Charles C. Shelton*

ATTORNEY

#150,263. Helicopter, July 13, 1948. This is the design patent for the S-51 helicopter. Design patents are used to register the overall shape of the design to prevent others from building a helicopter (or a model helicopter) with the same shape without the permission of the patent holder.

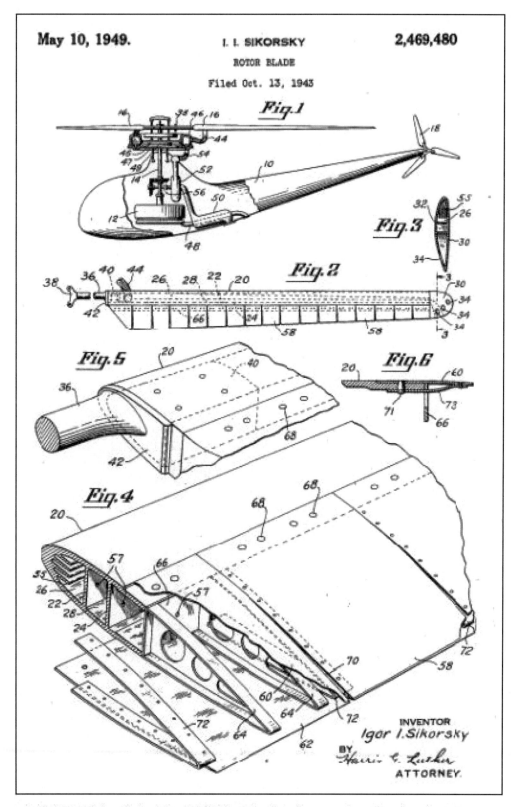

May 10, 1949.   I. I. SIKORSKY   2,469,480
                                   ROTOR BLADE
                              Filed Oct. 13, 1943

Fig.1

Fig.2

Fig.3

Fig.5

Fig.6

Fig.4

INVENTOR
Igor I. Sikorsky
BY
Harris C. Luther
ATTORNEY.

#2,469,480, Rotor Blade, May 10, 1949 - Patents the long-running Sikorsky rotor blade con-struction of an extruded metal blade spar with attached trailing edge pockets, as used on a number of generations of Sikorsky helicopters.

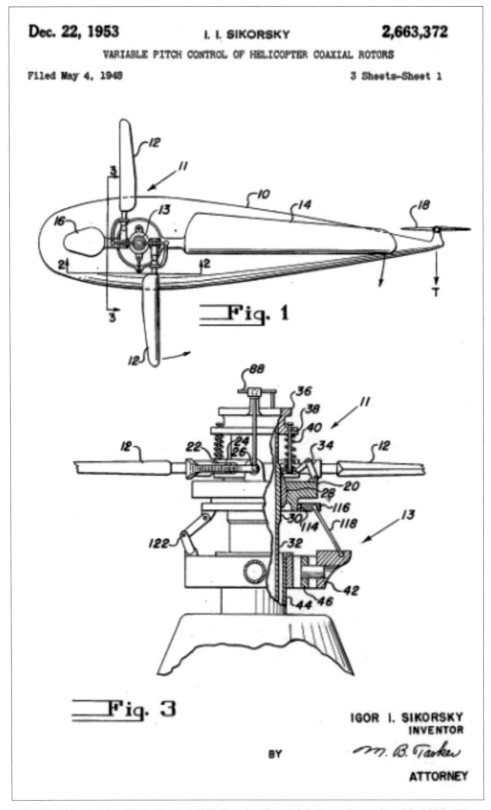

#2,663,372, Variable Pitch Control of Helicopter Coaxial Rotors, December 22, 1953: The ability of the pilot to directly control ever-larger rotors had become an issue and this patent proposes direct pilot control of a smaller rotor within a smaller rotor. Although not implemented, it does show that Igor Sikorsky never lost sight of the use of a coaxial rotor system.

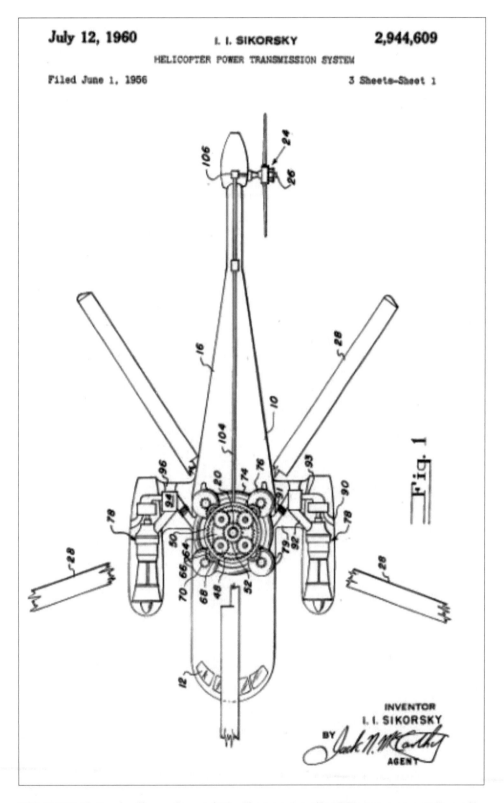

July 12, 1960   I. I. SIKORSKY   2,944,609

HELICOPTER POWER TRANSMISSION SYSTEM

Filed June 1, 1956   3 Sheets-Sheet 1

Fig. 1

INVENTOR
I. I. SIKORSKY
BY *Jack N. McCarthy*
AGENT

#2,944,609, Helicopter Power Transmission System, July 12, 1960: Igor was wrestling with the transmission of the engine's power to the rotor in increasingly larger helicopters. This patent proposes a system that would eliminate much of the intermediate gearing and shafting. This has the additional benefit of giving the designer more flexibility in the location of the engines within the aircraft.

Igor Sikorsky's Patents

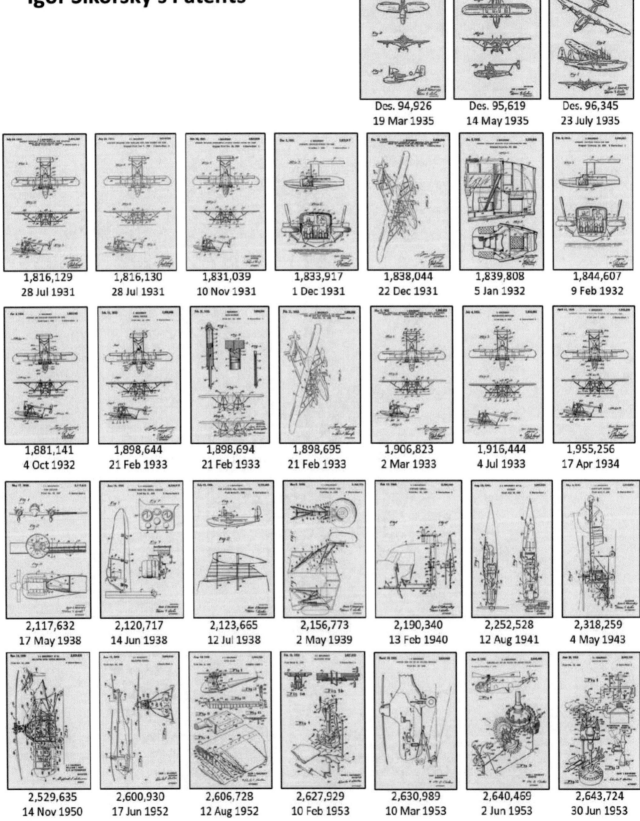

Des. 94,926
19 Mar 1935

Des. 95,619
14 May 1935

Des. 96,345
23 July 1935

1,816,129
28 Jul 1931

1,816,130
28 Jul 1931

1,831,039
10 Nov 1931

1,833,917
1 Dec 1931

1,838,044
22 Dec 1931

1,839,808
5 Jan 1932

1,844,607
9 Feb 1932

1,881,141
4 Oct 1932

1,898,644
21 Feb 1933

1,898,694
21 Feb 1933

1,898,695
21 Feb 1933

1,906,823
2 Mar 1933

1,916,444
4 Jul 1933

1,955,256
17 Apr 1934

2,117,632
17 May 1938

2,120,717
14 Jun 1938

2,123,665
12 Jul 1938

2,156,773
2 May 1939

2,190,340
13 Feb 1940

2,252,528
12 Aug 1941

2,318,259
4 May 1943

2,529,635
14 Nov 1950

2,600,930
17 Jun 1952

2,606,728
12 Aug 1952

2,627,929
10 Feb 1953

2,630,989
10 Mar 1953

2,640,469
2 Jun 1953

2,643,724
30 Jun 1953

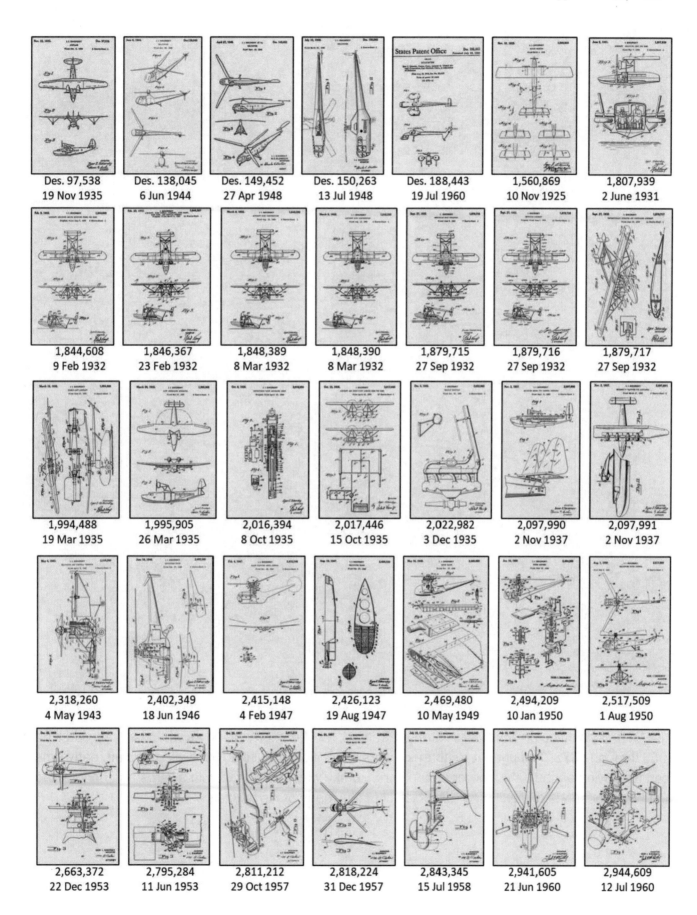

Des. 97,538	Des. 138,045	Des. 149,452	Des. 150,263	Des. 188,443	1,560,869	1,807,939
19 Nov 1935	6 Jun 1944	27 Apr 1948	13 Jul 1948	19 Jul 1960	10 Nov 1925	2 June 1931
1,844,608	1,846,367	1,848,389	1,848,390	1,879,715	1,879,716	1,879,717
9 Feb 1932	23 Feb 1932	8 Mar 1932	8 Mar 1932	27 Sep 1932	27 Sep 1932	27 Sep 1932
1,994,488	1,995,905	2,016,394	2,017,446	2,022,982	2,097,990	2,097,991
19 Mar 1935	26 Mar 1935	8 Oct 1935	15 Oct 1935	3 Dec 1935	2 Nov 1937	2 Nov 1937
2,318,260	2,402,349	2,415,148	2,426,123	2,469,480	2,494,209	2,517,509
4 May 1943	18 Jun 1946	4 Feb 1947	19 Aug 1947	10 May 1949	10 Jan 1950	1 Aug 1950
2,663,372	2,795,284	2,811,212	2,818,224	2,843,345	2,941,605	2,944,609
22 Dec 1953	11 Jun 1953	29 Oct 1957	31 Dec 1957	15 Jul 1958	21 Jun 1960	12 Jul 1960

Igor Sikorsky's Office

Igor Sikorsky's office on North Main Street, Stratford, has been preserved as when he occupied it from 1956 through 1972. Throughout his three careers in aviation -- Russian fixed wing aircraft, early American fixed wing aircraft. and helicopters -- Igor Sikorsky was recognized with awards and honors and met with thousands of dignitaries. His office contains many of his artifacts.

(10.) Igor Sikorsky's desk contained a library of books including four he authored (The Story of the Winged-S, The Evolution of the Soul, The Invisible Encounter and The Message of the Lord's Prayer). Several editions of the Bible and aeronautical reference books, his slide rule and models of the S-43 and S-38 aircraft are also on the desk.

(11.) Igor Sikorsky's fedora (not shown in above picture). During the Korean war the helicopter proved itself as a lifesaving aircraft and thousands of lives were saved. Around that time, word spread within the U.S. Marine Corps that if a pilot wore the Sikorsky fedora for just a few seconds, he would never be hurt while flying a helicopter. This special protection became legend, and Igor Sikorsky made certain that the hat would be readily available to his aviator visitors.

Russian/American Fixed Wing Memorabilia

2. On December 18, 1925 Igor Sikorsky, returning to wooded Long Island, made a hasty dusk landing in his S-29A. The wing of the sesquiplane airliner severed a tree limb. A piece of the embedded tree was mounted on a plaque to remind him of his narrow escape.

1. Painting of Igor Sikorsky's S-21 (The Grand). On August 2, 1913 the Grand set a record by flying 1 hour and 54 minutes over St. Petersburg while carrying eight passengers.

4. A 1938 design concept, designated as the S-45, would carry 100 passengers nonstop from New York to London. The project was terminated.

3. Igor Sikorsky is pictured at the controls of his S-5. The photo was used when he was issued Federation Aeronautique International (FAI) Pilot's License No. 64, dated August 18, 1911.

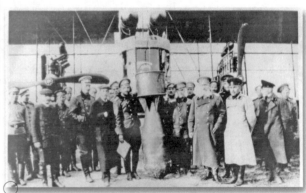

5. The Ilya Muromets (S-22). A group of officers of the "Squadron of Flying Ships" and a 1,000 pound bomb. Igor Sikorsky if fourth from the left and Russian Major General Shidlovsky is third from the right.

Early Rotary Wing Memorabilia

6.) On October 6, 1943, the VS-300 was donated to the Ford Museum in Dearborn, Michigan. (l to r) Henry Ford II, Charles Lindbergh, Sikorsky test pilot Lester Morris, Henry Ford and Igor Sikorsky.

7.) Igor Sikorsky and Charles Lindbergh on October 9, 1940. Mr. Lindbergh was a witness when, on that day, the VS-300 established an altitude record of 100 feet.

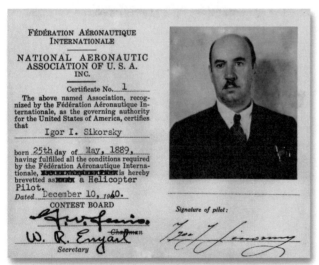

8.) Igor Sikorsky's helicopter license #1, awarded by the Federation Aeronautique International (FAI).

9.) Igor Sikorsky, Orville Wright and Colonel Frank Gregory meet at Wright Field, Dayton, Ohio. The Sikorsky XR-4 2-place helicopter was flown from Bridgeport, Ct. to Dayton for test flights and pilot training by the Army Air Force in May 1942.

Rotary Wing Recognition

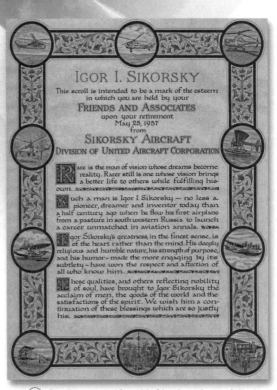

⑫ Igor Sikorsky met with a number of U.S. presidents, including Lyndon Johnson (above) and ⑬ Dwight D. Eisenhower (below).

⑭ In 1967 Igor Sikorsky was awarded the National Medal of Science, an honor bestowed by President Lyndon Johnson.

⑮ President Harry Truman awarded the Collier Trophy to the helicopter industry for its service in the Korean conflict, the military services and the Coast Guard on December 17, 1951. Mr. Sikorsky represented the helicopter industry.

⑯ Scroll presented from his friends and associates to Mr. Sikorsky at his retirement party – May 1957.

Astronaut / Test Pilot Memorabilia

(18) *Igor Sikorsky presents models of the S-61 (SH-3) recovery heli-copter to Apollo 16 astronauts John Young, Charles Duke and Thomas Mattingly in May 1972. Connecticut Senator Lowell Weicker (right background) looks on.*

(17) *The first man on the moon, Apollo astronaut Neil Armstrong, meets Igor Sikorsky in November 1970. Inscription from Armstrong reads "To Igor Sikorsky – With Admiration and Respect of a junior exponent of Vertical Take off & Landing."*

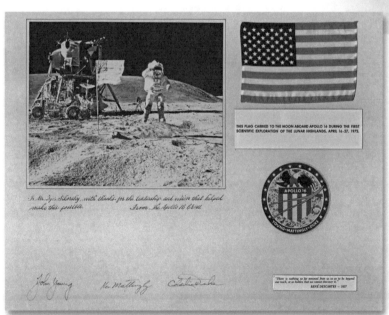

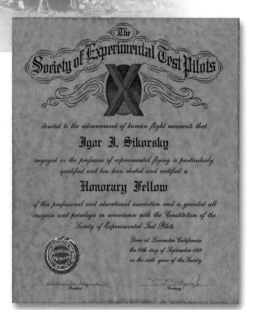

(20) *The prestigious Society of Experimental Test Pilots awarded Igor Sikorsky an Honorary Fellow award on September 29, 1962.*

(19) *A plaque from the Apollo 16 crew to Igor Sikorsky contains an American flag carried to the moon. The handwritten message states "To Mr. Igor Sikorsky, with thanks for the leadership and vision that helped make this possible."*

Igor Sikorsky's Last Letter

While on summer vacation with his father in Germany in 1908, Igor Sikorsky learned the details about the successful flights of the Wright Brothers. Following the trip, Sikorsky made the firm decision that aviation would be his life work. His entrance would be by way of the helicopter, a vehicle that could hover in one spot, rise and land vertically and fly in any direction chosen by the pilot.

In February 1972, a horrific fire engulfed the 26 story Pirani building in Sao Paulo, Brazil. The fire swept up from a lower floor and trapped people on the roof. Twenty helicopters took part in the rescue operation; 400 were rescued.

Mr. Jerome Federer, Director of the Flight Safety Foundation, sent Mr. Sikorsky details of the fire and the lifesaving role helicopters played. Mr. Sikorsky's response confirmed that his life-saving vision 64 years earlier had come to fruition.

Sikorsky Aircraft DIVISION OF UNITED AIRCRAFT CORPORATION

STRATFORD, CONNECTICUT 06602

PHONE (203) 378-6361

October 25, 1972

Mr. Jerome Lederer
Flight Safety Foundation, Inc.
1800 N. Kent Street
Arlington, Virginia 22209

Dear Mr. Lederer:

Please accept my sincere thanks for your recent letter and for the enclosure describing the Sao Paulo helicopter rescues. I had it read to me (my eyesight has failed to such extent that I can no longer read) and found it interesting indeed.

I always believed that the helicopter would be an outstanding vehicle for the greatest variety of life-saving missions and now, near the close of my life, I have the satisfaction of knowing that this proved to be true.

It was good hearing from you and I thank you again for sending me this information.

With kindest personal regards.

Sincerely,

I.I.Sikorsky

IIS:ccs

21. The last letter dictated and signed by Mr. Sikorsky. He passed away the following morning, October 26, 1972 at the age of 83.

Recollections of a Pioneer

Oct. 24, 1968

Ladies and Gentlemen: It is a great privilege to address you this morning on the subject of my career which covers the past 60 years of my life. My start in aviation dates to the year 1908. It would be difficult for you perhaps to visualize today what kind of a task or occupation flying was at that time. It definitely was not yet an industry; it definitely was not yet a science. Perhaps more correctly, there were a number of scientific theories and various information which were often contradictory. A few theories were correct, but no one knew which were the correct ones; It was at the time when the flights of the pioneering Wright Brothers became known and accepted. As I am sure you know, the Wrights' first flights were on December 17th, 1903. However, the information was unknown in Europe and largely unknown even in America. It may seem strange to say, but in a newspaper printed in Europe in the English language some time after they had already flown - it must have been at least a year or so - there appeared an editorial entitled "Flyers or Liars". The text went on to say that when a prominent scientist - and the name was mentioned - with undisputable logic proved that human flight is impossible, why should the public be fooled by the silly stories of two obscure bicycle repairmen who haven't even been to college. That was the status of aviation. Competent scientists proved mathematically that it couldn't be done and yet some people, some enthusiasts, tried to push it ahead.

I personally started my aviation career in 1908 while vacationing at Pau, France. My decision made within 24 hours of the moment when at last, I read in the newspapers the report about the many flights of the Wright Brothers. Before I had read the report, I was enthusiastically interested in flying even as a kid, that I could feel more than know what was right. When I saw the description of the flights and saw the photographs, I knew this was authentic. This is true, but I was greatly surprised to note that the report was printed on the second or third page of the newspaper instead of being printed in big letters right on the front page. Well, that got me going. I started immediately to develop my first helicopter. By this time in the year 1908, exactly 60 years ago, I had on my drafting board the drawings of a helicopter. At the end of the year I went to Paris, which was the center of flying activities in Europe at that time. The object was to purchase an engine for this new aircraft of mine. When I built it, the aircraft could not fly but I could be excused. I must explain - and there were several other excuses - that the engine I purchased weighed 7 pounds per horsepower and even today would be too heavy for a good helicopter.

I went to Paris, and I really dived, so to say, into the new aviation which was neither science nor industry, but rather an art or a passion. It was an intensely interesting and I would say romantic period of flying. Almost every day I spent a few hours on one of the 2 flying fields near Paris, Issy de Moulineaux, near the Eiffel Tower and Juvisy, some 15 miles or so from Paris. Sometimes, although quite seldom, I saw flights reaching the altitude of this room. As a rule, the best operation was what the Frenchmen called "Rouler" - which means "just running along on the ground." This was the way the French pilots taught themselves to fly. Most of the airplanes really could not fly. It was of value for trying the controls, trying to keep the tail in the air and to maintain the desired heading of the aircraft while running forward on the wheels. That was the best training available at that time. The running was not done at a terrific speed, probably something less than 35 miles per hour. Aviation was very young and the amount of information available was extremely scarce. Frequently the available data was later proved to be wrong or superstitious. For instance, one theory maintained by many scientists at that time was that the wings should be thin with a sharp leading edge. For instance, the Wright Brothers' wings were very thin. The earlier Farman airplane wings were also very thin. In contrast, at that time Russian Professor Joukowski proved that for the relatively slow speed of that time, moderately thick and curved air foils would be very much better. However, the "common sense and logic" required the sharp thin airfoils, and everyone including myself was using them. Well enough of excuses, although it may be said that they would eventually be used in the supersonic aircraft, but the time was way too early for supersonic flight.

Some of these superstitions, or unknown factors, were unfortunately learned only by crashes and fatalities. For instance, at that time the available knowledge concerned balloons and birds. In both of them we could not suspect the possibility of negative loading or the fact that the air may be pressing the wing downward and not upward. Upward loading was obvious on the spherical balloon. You would never see the balloon pressing down and the cabin jumping up in the air. That was simple. Consequently, in the airplanes at that time as a rule, there was no strength at all built into the wings to accommodate the negative load factors developed in reverse loading. All that was needed, or at least supposed to be needed, was just to support the wings so they would not fall when the airplane was at rest on the ground. A significant fact was learned when a very famous French aviator of that time, Chavez, the first man to cross the Alps by an airplane, approached for a landing on the Italian side of the Alps. Just before landing in full view of a large crowd of observers; a downward gust of air hit the airplane. The observers could see that the wings were broken downward and not upward. He crashed and was killed. From then on negative load factors were acknowledged and the designers started to build in a certain amount of strength for reverse loading. There were a number of other ways the new art was learned, some unfortunately by catastrophes. Other information was gained from the study of birds. Some knowledge was developed from intuition. Sometimes the intuition was wrong causing personal misfortune without actual physical tragedy. I remember cases when a man, even middle-aged people, sometimes were so deeply influenced by their own ideas - so much convinced that what they did would be right and indisputably successful that they would spend their life savings, sell their homes, or borrow to the limit of their capacity, and then build something which had already been proved to be wrong. The result, of course, was a total and complete failure.

The failures in life were almost as tragic as the actual crashes. At that time there was about one fatality per every 2,000 miles flown. There was probably at least one big crash for every 100 miles flown. Now these were the conditions when I started my career as a 20-year-old youngster. I hoped to build a successful helicopter. At that time, it was considered to be impossible and a false way of even approaching flying. I will continue the story of my personal efforts by reference to a number of illustrations. I will do it not because I attribute any particular importance to them, but I want to show you and to speak to you gentlemen about some aircraft of which perhaps I am one of the very few who can speak. Most of the other aircraft could only be studied in historical papers. The only undisputable qualification for which I can claim the honor of addressing you is that I am still alive and most other designers are not.

Figure No. 1 shows my very first helicopter built in 1909. It gave me important information, training and

Figure 1. H-1

Figure 2. H-2

intuition with respect to handling engines and structural factors pertaining to aviation. I soon learned that it would not fly. Therefore, I constructed a set of large scales. With the helicopter mounted on the scales I measured the lifting capacity of the ship. The lifting capacity was found to be about 300 pounds which was not really bad with an engine of 25 HP. But the helicopter weighed about 400 pounds, so that was that.

Figure No. 2 shows my helicopter No. 2 built in 1910. I learned more about aircraft construction. I was able to reduce the weight of this ship to about 300 pounds and to increase the lifting capacity to about 350 pounds. Consequently, this ship could at least definitely lift its own weight. Meanwhile, more by feeling and intuition than otherwise, I learned that there was a multitude of other problems beyond having sufficient lift to permit flight. Consequently, while the second helicopter was still under construction and test, I started to build my first airplane.

The first airplane also could not fly because it was powered only by an engine of 15 HP. It could run over the ground pretty slowly, probably not over 25 or 30 miles per hour, and it gave me basic training about controlling an airplane. When I decided to discontinue work with the helicopter, the 25 HP engine from the helicopter was installed in the airplane, with various changes to produce what I knew as the S-2, the second airplane. On June 3, 1910, I made my first solo flight. The flight was not a very long one, although it might have been about 600 feet. It was probably up to 3 or 4 feet high but it was a real flight. That was the beginning of my personal flying. It was also the very first flight which I ever had in my life in anything. I had never been a passenger in the air in anything, not even in a balloon, not to say an airplane. I had to follow the very interesting, even though difficult, career of an early pioneer to build an aircraft without knowing how to build it, then climb into the pilot's seat and try to fly it without knowing how to fly it. When a crack-up occurred, which happened very very frequently, I did not know whether to blame the pilot or designer or the builder. In a way this was convenient for it eliminated any quarrels or disagreements. The guilty party was well-known and the only thing was to get started again and rebuild the aircraft.

Figure 3. S-3

Figure No. 3 is my airplane No. 3. As an indication of the time, I bring your attention to the automobile which was an open car and on which there was no windshield because the average speed of the automobile was such that a windshield was not yet necessary although the riders usually wore goggles. No. 3 was substantially better than 2 but it crashed after it had been in the air less than 10 minutes.

So I became much more careful and produced airplane No. 5 with a more powerful engine, Figure No. 4. I planned to train myself from the very beginning. This was early in my third year of aviation activity. What I reliably learned during those 3 years was the degree of my ignorance in this business, and in fact, the ignorance not only of myself but most everybody else. Having learned this, I also learned something else. I knew I was more afraid, not of personal injury, but I was afraid of cracking the airplane which would· mean another half year of hard work of rebuilding. This would mean coming back to my family and asking for more money which was not amply available. My family was helping me at that time, so I tried to be careful, and I planned with this airplane to do the following: I would fly for a long period only in a straight line. There were no airports at that time. There were only. fields with grass and that was all-pastureland--so I could fly straight and land at the other end, roughly

Figure 4. S-5

3/4 of a mile distant or even less. If I were to go beyond, there might be a ravine or an occupied and built-up area. In other words, if I went beyond the edge of the meadow, I would have to make the whole circle, with sufficient confidence to come back to the same spot. That would be much more difficult. Probably if there had been a real airport, I could have taught myself easier with fewer crackups and difficulties. On this ship I decided that for a long while I was going to fly in a straight line against the wind and nothing more, landing at the other end of the field. Then I would have a peasant with a horse tow the airplane back to the point of departure and I would do the same thing again. I repeated this a dozen or more times a day, and day after day, training myself until at least I became familiar with this particular maneuver. I could make slight deviation from a straight line; in other words, slight turn left, follow it by a turn right to come back to straighten out and to land at the end. Finally, I decided that I knew enough, and I flew over the ravine, flew over the inconvenient places, circled in the air, and landed back near my starting point after a 3 1/2-minute-long flight, which was a significant event in my life. For the first time I had landed when I wanted to and not just when there happened to be a landing. My further progress went fast and by the middle of the year 1911 I obtained my International Pilot's license from the Imperial Aero Club of Russia. It was No. 64 indicating it was still a pretty early time in aviation.

My dream for a long time was to construct a big airplane with several engines. At that time the stopping of the engines in flight was one of the typical troubles of aviation. So, I dreamed about a ship having several engines and a cabin which was particularly necessary for the severe Russian winter climate. Now finally an airplane to my mind had to be like a ship piloted or manned by a whole crew and not one man: one man or even preferably two, to be a pilot, one to be a mechanic and one to be a navigator. This is a standard arrangement at the present time. At that time, it was a complete novelty. Such a thing as a closed cabin for the pilot was considered not merely a novelty but was considered an impossibility. There were no flight instruments whatsoever at that time, so in building this airplane, I had to imagine some sort of flight instrument. I had virtually three and these proved to be adequate. One was at a curved tube with a steel ball running inside through some liquid. This was obviously not a bank and turn

Figure 5. S-21 "The Grand"

instrument but rather an indicator that the ship has correct position in spite of the turn, so whether you fly straight or you turn, even very sharply, you have to keep the ball in the middle and that would be the safe position. That would be good for the time being. Next, of course was the altimeter, which was available and finally, there was the speed indicator. This was one which I had to create myself. I took a U-shaped tube of glass with a little alcohol inside, colored to make it visible, and I connected one end to a sort of receiver of pressure. I calibrated the instrument in accord with what were the then recently available information about speed and air pressure. While the airspeed indicator proved to be not exactly correct, it always showed the right relationship. Consequently, I was able to learn exactly what was my indicated take-off speed and what was my indicated landing

speed and what was my indicated cruising speed, and consequently this was all I needed. Outside of that I had to train myself for the new condition of the pilot who flies without the familiar stream of air in the face, without the familiar sensation of greater noise in one ear than in the other which would indicate side slip, etc. This ship, "The Grand", was built in 1913, Figure No. 5. It was the first 4-engine airplane to really fly - successfully.

Figure No. 6 shows the visit to this airplane by the late Czar Nicholas of Russia in the summer of 1913.

Figure 6. Emperor Nicholas and Igor Sikorsky

Figure No. 7 is a closer view of the Emperor Nicholas and myself on the balcony of this airplane in the summer of 1913. This airplane known as the "Russian Knight" or by the familiar popularly given name "Grand" from the French word "The Big One", was the forerunner. The first ship was very much overweight. The four 100 HP engines could barely carry it, but a most important result was proved. The main part of the criticism of multi-engine airplanes was that it would not be possible to fly a ship with one dead engine because it would lose control. It was contended that 4 engines would prove even more dangerous than one because the loss of an engine on one side would cause a crash. With the twin tandem en-

Figure 7. On the S-21 Balcony

gine "Grand" it was proved that this was not the case. With adequate design and placement of rudders it was possible to counteract the wrong torque of the engine thrust and to keep on flying. On this ship, even though two engines on one side were stopped, the ship would not lose altitude and the control characteristics were satisfactory. Consequently, it was possible to immediately sell the idea to build the next ship.

Before we go to the next ship, I would like to tell you a brief strictly personal story. I had been dreaming about a big multiple engine, usually a four-engine ship, for a long time. In 1912 I became a member of a fine big organization, Russian Baltic Car Factory, which at that time had at the head as its president, an

Figure 8. S-6

extremely fine progressive and intelligent man whom everyone respected. I liked him quite well, and even while a youngster, I had visited him in his home from time to time because he was deeply interested in flying. Now at that time, in 1912, there was a military competition in St. Petersburg in which I participated with my single engine airplane, the S-6, Figure No. 8. On September 17, 1912, I was invited after one flight of the competition to dine with the head of the Russian Baltic Car Factory, which was a great honor for me, and so I went there. After dinner we both -- I mean my host and myself -- went to his studio to drink black coffee which both of us liked and to talk about the flying competition. My host proved not to be interested in the progress of the competition and consequently, the discussion shifted to another subject. Another subject was introduced, and I said a few words about a large aircraft with several engines, with a crew, with a closed cabin. At first, I briefly described the characteristics of a ship which should be built some time. I stopped a few times, but my host always said, "Go ahead, go ahead". So, I mentioned more and more details and finally described the whole project. About midnight I started to go home and completed my discussion saying: "Suppose we win the competition, let's invest the money in the construction of the first big multiple engine airplane." The prize was the equivalent of $30,000 which was a substantial amount of money for that time. My host firmly said: "No, begin the construction immediately." That was midnight of September 17, 1912. It was probably a decision which may be directly responsible for my fact of addressing you now here. I went home. My home was situated near the small factory in which we were building out of wood all the single engine airplanes of mine. I woke up the policeman in charge, asked him to immediately get hold of my two foremen, my Chief Engineer, and two or three other men. Our Engineering Department was not exactly as big as our present one but there may have been a half dozen men or so. It was about 1:30 in the morning but the men, surprised, sleepy and bewildered assembled in my living room. I served them a glass of wine and then I reported the great news. It was accepted with the greatest enthusiasm. In fact, at that time in aviation to be successful one had to be enthusiastic. It was not - yet a job - not yet a position. It was a passion. Now for the next 2 or 3 hours I had one of the most successful engineering conferences in my life. The job was difficult, we knew that. It was difficult for another reason. It required totally new engineering and there were a number of difficulties with respect to materials. There were no large strong wheels, so we decided to use 16 small wheels combined in pairs of 2 forming two under-carriages of 8 wheels each. There were no high-grade steel wires available except in the actual piano factories where they were building pianos for music. They we were able to get the strong wire. The small steel tubes we had to get from a bicycle shop.

Figure 9. S-22

Now next day the job was started. This was September 1912 and on May 13, 1913 I made the first flight in this airplane. It flew with such success that it resulted in the construction of the next larger and very much superior aircraft. It flew with such success that it resulted in the construction of the next larger and very much superior aircraft.

This is the next aircraft that is, perhaps, the truly successful forerunner of the 4-engine commercial and military airplanes, Figure No. 9. This ship was built in the latter part of 1913 and I tested it out very early in 1914 before World War I. The early ships still had the 100 HP engines, which were later replaced with 140 HP engines. The wings had a bigger span. It had a gross weight of approximately 5 tons, and it could fly very much better in every respect than the previous one.

Figure No. 10 shows the world record of the number of passengers which I established in February 1914. The prior record belonged to Louis Bleriot who carried 10 people. I carried 15, in other words, a total of 16 on this ship, and this is a

Figure 10. World record for passengers carried

photograph of my passengers - 16 men and one dog. Now I bring your attention to the balcony on the top of the fuselage. The cruising speed of this ship was perhaps of the order of 70 miles 'per hour. Now a wind of 70 miles, if perfectly smooth as it is in flight, is not too objectionable. So, I could stand on this platform, and I did so a number of times. I. found it probably the most pleasant and most inspiring sensation I ever had in flight. The balcony was way behind the wing. The view from there was miraculously interesting and beautiful, particularly when flying around or above the clouds.

In Figure No. 11 you can see the ship, "Ilya Mouramets", -- that was the name of this aircraft and all the following aircraft -- approaching for a landing, and you can see 2 men standing on this platform. I may state that, of course, I had to train myself to fly the big ships. Obviously, being a designer, I wanted also to be the test pilot, which I was. As long as the ships were so expensive and certainly not expendable, having once learned to fly, I didn't let anyone else land the ship. I did all the landing even though in the air a number of other pilots flew them. So, I was the only pilot to take off and land 4-engine airplanes from May 1913 and until the late fall 1914 when I trained and checked out the first group of Army aviators.

Figure 11. S-22 "Ilya Mouramets"

Figure 12. S-29A

Now we turn to my American experience. Figure No. 12 shows the S29A, the very first airplane which the young Sikorsky Aero Engineering Corporation started to build in 1923 and completed in 1924. It had 2 Liberty engines and may have been one of the first, if not the first, twin engine ship capable of flying with one engine dead. That's myself sitting at the controls of this ship. I was still undecided as to whether the rear position of the pilot or front position would be preferable. The rear position had advantage of giving a good

sight of the whole aircraft in flight. Furthermore, it was more familiar to the pilots since all single-engine airplanes at that time had the pilot sitting in the rear. Later we abandoned it for the front position which certainly offered better visibility of the surroundings.

Now from the S-29A airplane we move on to the S-38 amphibian airplane, Figure No. 13. Several ships had been produced in between, but this one proved to be the real originator of this company. For the

Figure 13. S-38

first time we were able to sell a reasonably large number of these ships. The total sales amounted to several million dollars which was stupendous at that time. It was this ship that virtually created this organization in America. It was extensively used by several airlines, in particular by Pan American Airways, to pioneer their Caribbean, Central and South American routes.

On the basis of our information and established reputation, we got from Pan American Airways an order for the first really large passenger carrying airplane, the S-40, Figure No. 14. The largest ships up to that time were built by Fokker and Ford. Each of them carried less than 20 passengers and this new ship had to carry 40 passengers. It was built for Pan American Airways and proved to be a very successful aircraft. I had enough experience to realize that the large size of the ship would present several major difficulties. I thought that it would be better not to augment the difficulties by making a more modern looking airplane even though advanced design was known to us. For example, we could have placed the engines in a better streamlined

Figure 14. S-40

way. Instead, I used the old-fashioned formula following the configuration of a ship that was successful - the S-38. The large S-40 was very controllable, very usable. Three ships were sold to Pan American Airways and were extensively used for a whole decade. Not a single one was cracked up and they were all retired for obsolescence at the beginning of World War II.

During the first flight of the S-40 with passengers from Miami to the Panama Canal, Charles Lindbergh was the chief pilot and the actual director of this flight, figure No. 15. This is Lindbergh at the controls of the S-40. The rest beside the crew were passengers. The ship ·was full to capacity and I was among the passengers although I spent my time mostly in the pilot's cabin. Every evening I dined with the crew in some different spot. During these dinners Lindbergh and I were usually together. On the backs of

Figure 15. Charles Lindhbergh and Igor Sikorsky in the S-42

the menus of the restaurants we would write the specifications and make sketches for a ship capable of flying across the ocean. The flight to Panama was a great success. I became a good life-long friend with Lindbergh whom I consider to be an outstandingly brilliant pilot and citizen of this great country. Following the return flight, we departed with the idea that he, Lindbergh, who was then the consultant of Pan American Airways, and I as Engineering Manager of this organization then, and in fact for a whole stretch of 34 years, were to promote the transoceanic airplane. I was going to sell the idea to United Aircraft to accept a contract, and Lindbergh was to sell to Pan American the idea of giving United Aircraft a contract for the design and construction of a transoceanic flying boat.

Now this Figure No. 16 is the result of this dream, the S-42, which proved to be a very advanced design flying boat. At one time this single aircraft held 10 records for flying boat or hydro airplanes out of a total of 20 world records of this category. In other words, this ship alone held 10 world records, while the other 10 records were divided among all other flying boats or hydro airplanes of all other nations. The

Figure 16. S-42

S-42 pioneered commercial flying over the Pacific Ocean in 1935 arid over the Atlantic Ocean in 1937. This schedule calls for an explanation. The flight over the Atlantic had to be in accord with the British schedule of what then was called Imperial Airways. They were not ready to send their boat from England across the ocean to the United States until 1937. Consequently, Pan American out of courtesy and agreement would not send an American boat to England. In 1937 the S-42 starting from America, and the Short Brothers flying boat starting from England, crossed the Atlantic at the same time. This was the beginning of regular transoceanic passenger-carrying flights. Between 1935 and 1937 the opportunity was utilized to develop the flying and navigational techniques with pioneering flights over the Pacific Ocean to the Hawaiian Islands, Asia, Australia, and New Zealand.

Figure 17. S-44

The conclusion of our flying boat wor was accomplished by the S-44 flying boat, which you can see here, Figure No. 17. This ship was built for a special patrol and bombing objective for World War II. The basic objective was to fly non-stop across the Atlantic and similar distances in the Pacific area. The previous airplane, the S-42, while flying across the Atlantic had to stop either in Bermuda and the Azores, or in Newfoundland and Iceland, because the non-stop distance was too far. With the S-44 airplane it was possible to fly non-stop between New York and London.

Our flying boat career was entirely successful and there were subsequent opportunities to design and-construct even larger flying boats based on our reputation for building good flying boats. However, it was a limited market by today's standards and large airports to accommodate landplanes were being built throughout the globe. This competitive situation prompted renewed consideration of the helicopter.

As the helicopter project was discussed with my associates at United Aircraft, caution was expressed for my reputation as a good designer of large airplanes and flying boats. There was concern that it would not work, or if it did, nobody would need it. For instance, I was asked if I knew when the helicopter would fly as fast as the airplane. My reply was, "Of course I know. The answer is - Never." "Do I know when the helicopter would be at least more efficient than the airplane in load-carrying." My reply was again "never", but the helicopter can do things which no airplane and no aircraft and in fact, no other vehicle ever used or dreamed by man can do. This was the reason why in 1938 I went to the management of United Aircraft to the late Mr. Brown, then president, and Mr. Wilson and some other officials of

the organization. I sold the idea that United must go ahead on its own with the building of helicopters. I was confident at that time that I would be able to make a helicopter which would fly.

Now this is our first VS-300 helicopter, Figure No. 18, and one of my first flights. Once again, I was the test pilot of the machine, and this is one of the first flights conducted in the late fall of 1939.

The longest flight in 1939 lasted about 2 minutes but we were certainly flying. In the next year, 1940, the flights became longer and more controllable. Here in Figure No. 19 is

Figure 18. VS-300

one of my flights deliberately hovering in front of a big tree to show the exceptional hovering and control characteristics of the aircraft.

The progress continued. In 1940 I was able to stay in the air for 15 minutes and by May of the next year (1941) I established the world record for endurance at 1 hour and 33 minutes, Figure No. 20. At that time this was the longest recorded flight time of any helicopter in the world. That gives you the idea of how young the art really is.

Figure 19. VS-300 in hover

Figure 20. Endurance record

Figure 21. Changing a tire while in hover

At that time I firmly realized, as I continue to realize now, that the very first characteristic required of the helicopter is perfect control. The best and highest and most wonderful performance characteristics would kill the success of the helicopter if it did not have perfect control. Under what condition we will say a few words later. Now to really train ourselves and demonstrate such control, we started to invent a number of various, what we called, tricks. Now this is one of them, Figure No. 21. You can see here the helicopter hovering and a man taking off the wheel. The man took off the wheel, supposedly changed the tire, and then came back and put it in place while I was still hovering nearby.

Now this is another of the tricks a helicopter has called "spearing the loop." (Figure No. 22) That means flying with this stick projecting out in front of the helicopter and piercing the ring which was approximately one foot in diameter. Now in this case you can see that the pilot made the error of less than 2" in spearing the center of the loop. This and several other things of this nature permitted us first to train ourselves as pilots and to demonstrate the superb control characteristics of the helicopter.

Another very significant demostration is represented here, Figure No. 23. The ship is the XR-4 helicopter. The XR-4 was the forerunner of the production helicopter of which over 100 were sold. It was the very first really successful production helicopter in the world.

By contract, the very first helicopter had to be delivered to the U. S. Army Air Corps at Wright-Patterson Air Force Base in Dayton, Ohio. It was delivered there by air. Our Chief Pilot, Les Morris, was the senior pilot and I acted as the second pilot on a portion

Figure 22. "Spearing the loop"

of this delivery flight which was intensely interesting. The XR-4 was flown on the first cross country flight ever made in the United States or anywhere in the Western Hemisphere by helicopter. The photograph shows this helicopter upon its arrival at Wright Field in Dayton with myself at the left, Orville Wright who visited to see the machine in the middle, and General Gregory, then a Colonel, who was very instrumental in the whole procedure of the creation of the helicopter industry.

Figure 23. Igor Sikorsky, Orville Wright and Col Gregory

Ladies and gentlemen, it was a privilege to say these few words, to you today. I congratulate you in having selected a life of activity which to my mind is one of the most promising. I am convinced that the helicopter is here to stay for an indefinite period of time just as the automobile is here to stay and just as the high speed near sonic and supersonic liner is here to stay. The liner will never replace the helicopter and the helicopter also will never replace the liner. Although there may be an important role for the so-called VTOL aircraft, it would have to combine as much as practical and possible the characteristics of the pure helicopter with the high speed of the airplane. However, it must be recognized that you can never expect to have a radical VTOL design without losing the good hovering characteristics of the helicopter.

There are a number of types which have been proposed. Some of them very promising such as the compound helicopter and the stowed rotor helicopter. Nevertheless, they will never replace the pure helicopter and not replace the pure airplane. Each of those will stay in their own place. But the role of the helicopter, the jobs it can perform, jobs which nothing but a helicopter can do - are here to stay. They are certain to assure the importance and seriousness and permanence of the helicopter industry. As one who is just about ready to conclude both the professional and any other activities as well, I would like to complete my remarks by telling you about one of the things which remain a source of great personal satisfaction: The first thing is that a couple of years ago a good friend of mine, a French pioneer by the name of Col. Dolfus, director of the French National Museum of Aviation in Paris, sent me a brochure which was dated 1862 - not 19 -- but 1862 which consequently is now over 100 years old. The brochure speaks of the helicopter as "L'Aeronef, Appareil de Sauvetage", which means "a flying machine, an instrument for saving life." I had always hoped that the helicopter would prove to be such an instrument, and it did. Our helicopters have to the best of my present estimation saved over 50,000 lives which is a mighty good record. The task of picking up most of our astronauts from the sea was done mostly by our helicopters. The mission of carrying the 3 latest presidents has also been done regularly by our helicopters. **But the main thing, I would say, was the saving of life is a source of great satisfaction to me. I sincerely hope that in addition to a big success in general for yourselves, that the saving of life would remain the source of satisfaction and great interest to you who are entering the helicopter field.** I thank you.

Acknowledgements

The contents of this book are drawn from the Igor I. Sikorsky Historical Archives collection. The categorization, preservation, and analysis of the data are the result of the work of our dedicated volunteers. This book would not have been possible but for Archivists Vaughan Askue, Bill Bergan, Bob Blackwell, Hank Chmielewski, Ed Covill, Jose Dantas, Dan Godlewski, Andreanne Johnson, Art Linden, Ed Sullivan, and Paul Swanson.

Starting with the creator of the original 'Winged S' logo designed by Andrei Avinoff, Sikorsky has employed numerous artists, and we have used their works throughout this book. Special thanks to Archives volunteer Joe Keogan and his graphic design artist daughter Jodi Buckley. Joe's aircraft line drawings are used extensively, and Jodi has been invaluable in creating the layout for this book. Thanks also for the permission from Two Roads Brewery to use artist Ryan Pencast's paintings of Igor Sikorsky. Two Roads commissioned Ryan to produce these paintings for the labels on their specialty brew, Igor's Dream.

Special thanks must be extended to Lee Jacobson for his contribution of years of authoring our Archives News publication. We used his work as the basis for the Igor Sikorsky the Man section.

This book contains a large amount of detailed information on Igor Sikorsky, his company and his products. All data is from our Archives collection and would not have been accessable if not for the relentless efforts of our volunteer archivists in the categorization and preservation of our collection. Our major individual contributor has been Bill Paul, past president of Sikorsky Aircraft. We would not have been able to perform the research for this book if it were not with his support.

Finally, this book would not have been possible without Gene Buckley and other executives of Sikorsky Aircraft who recognized the historic significance of the memorabilia of Igor Sikorsky and his legacy.

This volume has been produced with the support of Sikorsky Aircraft, a Lockheed Martin Company.

About the Authors

Frank Colucci

Career journalist Frank Colucci has written in depth about the rotorcraft industry for over 40 years. The author of an early book on the AH-64 Apache, Mr. Colucci routinely writes about military and civil helicopter operations, cockpit and weapons system integration, aerostructures, flight simulation, and other rotorcraft topics. He was the North American Editor of an international military helicopter magazine and a regular contributor to numerous defense and aerospace industry publications. Mr. Colucci is currently a Senior Contributing Editor for Vertiflite, the magazine of the Vertical Flight Society, and he writes the quarterly newsletter of the Igor I. Sikorsky Historical Archives. He has received the Helicopter Association International Excellence in Communications Award and the Defense Media Outstanding Achievement Award sponsored by Lockheed Martin.

Frank Colucci holds of Bachelor's degree from Brooklyn College of the City University of New York and is a long-time member of the Vertical Flight Society (American Helicopter Society) and the Brooklyn Plastic Modelers Society. He is currently on the Sikorsky Archives Board of Directors.

John Bulakowski

John Bulakowski joined Sikorsky Aircraft in 1970 as a design analytic engineer and broadened a 32-year career with successive positions in engineering, marketing, licensing and program management where he ascended to Director of the U.S. Army / Air Force H-60 Product Line. John was part of the original Sikorsky design and proposal effort that won the Utility Tactical Transport Aircraft System (UT-TAS) competition for the Black Hawk helicopter. He later led the effort to build the Black Hawk under license in the Republic of Korea. Upon retiring, John continued as a contractor with Sikorsky for another 18 years in program management and Sikorsky proposal efforts for the U.S. Presidential helicopter, Marine Corps heavy lift helicopter, Air Force combat rescue helicopter and major international sales efforts. He currently serves on the Board of Directors and is Vice President of the Igor I. Sikorsky Historical Archives, Inc.

John Bulakowski holds Bachelor and Master of Science degrees in Mechanical Engineering and a Master's degree in Business Administration. He is a member of the ASME (American Society of Mechanical Engineers), AAAA (Army Aviation Association of America), and the VFS (Vertical Flight Society).

Igor I. Sikorsky Historical Archives

This book celebrates 100 years of aviation innovation, from the start of Sikorsky Aero Engineering Corporation in 1923 to Sikorsky, a Lockheed Martin Company, today. It is the work of the Igor I. Sikorsky Historical Archives. The Archives was founded in 1995 to protect and make accessible for research and education the documents, photographs, engineering drawings, correspondence, and memorabilia of Igor Sikorsky and his legacy in fixed- and rotary-wing flight. The Archives has been recognized as "one of the most important resources on the history of American aviation and technology in North America" and a "National Treasure" by former curators of The Smithsonian Institution and Yale University, respectively. The strength of the Archives is in the depth of our volunteer experience. More information on Igor Sikorsky, Sikorsky Aircraft and how to volunteer and contribute to the Sikorsky Archives can be found at our website -- https://www.sikorskyarchives.com.

100th Anniversary Edition Version 3 (11/9/22)

Made in United States
Troutdale, OR
12/12/2023

15769557R00115